Lubricants and Waxes

This book covers chemistry of lubricants and waxes up to technologies applied for their commercial production including technologies in commercial operation. Grease as a lubricant in solid form including various formulations characterizing grease and liquid lubricants has been extensively covered. The environment impacts of the processes along with environment friendly processes developed have also been covered along with data in the book. To make the book more practical, various plants' operating conditions have been provided as a part of case studies.

Features:

- Provides updated information on the process, technologies, and application especially for production of Group-II and III base oils.
- Covers theory behind the processes including stoichiometry of chemical reactions, physical and chemical structures, and theoretical text explanations.
- Includes mathematical equations and data to evaluate the actual operations and/or check design of the plant.
- Explores practical applications including commercial production of lubricants and waxes.
- Covers various industrial case studies.

This book is aimed at Professionals in industries involving the application of lubricants and waxes.

Lubricants and Waxes

From Basics to Applications

Dhananjoy Ghosh

CRC Press is an imprint of the
Taylor & Francis Group, an **informa** business

First edition published 2024
by CRC Press
2385 NW Executive Center Drive, Suite 320, Boca Raton FL 33431

and by CRC Press
4 Park Square, Milton Park, Abingdon, Oxon, OX14 4RN

CRC Press is an imprint of Taylor & Francis Group, LLC

ISBN: 9781032369082 (hbk)
ISBN: 9781032369099 (pbk)
ISBN: 9781003334422 (ebk)

DOI: 10.1201/9781003334422

Typeset in Times
by Newgen Publishing UK

Dedication

The book is dedicated to the memories of my heavenly parents.

Contents

Preface....ix
About the Author....xi
Acknowledgments....xiii

Chapter 1 Preliminaries....1

1.1 Defining Lubricant and Wax....1
1.2 Chemistry....2
1.3 Technologies....7

Chapter 2 Lube Oil Base Stock (LOBS)....11

2.1 Physico-Chemical Properties of LOBS and Significance....11
2.2 Classifications and Chemical Composition of LOBS....16
2.3 Standards / Market Specifications of LOBS....20
2.4 Manufacturing Technologies for LOBS....21
2.4.1 Processes Followed for Group-I LOBS Production....21
2.4.2 Processes for Group-II and Group-III LOBS Production....63
2.5 Application of LOBS....77

Chapter 3 Lubricant....79

3.1 Lube Oil....79
3.1.1 Lube Oil Properties and Significance....79
3.1.2 Lube Oil Classifications, Compositions and Applications....85
3.1.3 Lube Oil Marketing Specifications....106
3.1.4 Processes Followed in Lube Oil Production / Blending....108
3.1.5 Branding and Marketing of Lube Oils....112
3.2 Grease....113
3.2.1 Properties and Specifications of Greases....113
3.2.2 Classification, Composition and Applications of Greases....116

Chapter 4 Wax....121

4.1 Properties and Significance....122
4.2 Classification of Wax....123
4.3 Marketing Specifications of Wax....124

4.4 Application of Wax 125
4.5 Manufacturing Technologies for Wax 131
4.5.1 Solvent Deoiling of Vacuum Distillate 132
4.5.2 Solvent Deoiling of Aromatic Extracted Distillate (Raffinate) 140
4.5.3 Solvent Deoiling of Oily Wax from Tank Bottom 142
4.5.4 Solvent Deoiling of Hydrocracker Bottom 143
4.5.5 Hydro Finishing of Deoiled Wax 143
4.6 Wax Derivatives and Application 149

Chapter 5 Energy Optimization in Dewaxing and Deoiling Process 151

Chapter 6 Tribology 163

Chapter 7 Environment Impact and Mitigation in the Use of Lubricants 166

7.1 Bio Lubricant 167
7.1.1 Perspective 167
7.1.2 Classification / Composition / Properties and Application / Commercialization 169
7.1.3 Processes for Production 176
7.2 Development in Automobile Industry to Reduce Lube Oil Consumption 177
7.3 Recovery of Lube Oil from Spent Lube Oil 179
7.4 Ionic Liquid as Sustainable Lubricant 182

Chapter 8 World Lubricant Market and Major Operating Base Oil Refineries 183

References 185
Index 187

Preface

With the objective of reaching out to readers starting from undergraduate to post graduate students in petroleum technologies and particularly in lubes and waxes, practicing engineers in lube technologies' operation, corporations having plants in lubes and waxes in their endeavor to improve profitability and sustainability, up to the researchers to find the references, the book contains all details including chemistry of lubes and waxes, technologies in application along with their latest development.

While designing the contents of the book, the chronology has been given focus to include all relevant sub elements on the subject in sequential orders to achieve a flawless completion of the book. In laying foundation on the manuscript preparation, highlighting fundamentals on all sections / sub-sections given top most priority, viz. physical significance of each property of lubes, classification of lubes and waxes, market relevance, technologies presentation, and their evolution in line with aspiring market needs like cost competitiveness, quality improvement along with alternatives, environment impact and mitigation. Mathematical models, chemical stoichiometry, facts and figures, and pictorial demonstration are the main fundamentals in all the processes of lubes and waxes. All these aspects have systematically been covered in the book.

Lubes and waxes are of the similar boiling range and as in their production the main feedstock source is mostly mineral based, i.e. crude oils, the Lube bearing crude oils contain wax also in their composition, but the crude oils not having the Lube potential can have wax potential. Due to the above scenario, discussion on lubes and waxes simultaneously find a common platform.

Grease, a solid form of lubricant has also been discussed in context. Also, to cover mitigation on environment impact in using lubes, alternative products and methodologies have been discussed to take the readers and practitioners to future directions. The subject, tribology, has also been covered in its relevance.

About the Author

Dhananjoy Ghosh is an industrial professional and visiting faculty in universities. He is a chemical engineer, did his graduation with honors in chemistry also; he holds a PhD in petroleum technology, all from Calcutta University, India. Since beginning, his focus was on innovation, troubleshooting and safety. He had a long stint of experience in Lubricants and Wax technologies and operation when he carried out lots of field research in this field. He had been in many refineries for safe commissioning of their projects in these two areas. At senior level, he also pursued his career in operating other petroleum technologies, and in this way, he became the expert of operation in all petroleum technologies. Due to his aptitude to innovation, he developed technology of a new process, 'Re-extraction of Aromatics from petroleum residual extract' in 1995 and got innovation award from NPMP (under petroleum ministry of India) in 2000. In 1996, he developed another technology, 'Simultaneous deoiling and dewaxing in a single unit to produce quality base oil and micro-crystalline wax at the same time' with the objective of integration and energy savings which got commercially implemented in 2002. He has authored two books, viz. 'Safety in Petroleum Industries' published by CRC Press and distributed by Taylor & Francis, and 'Petroleum Refineries – A Troubleshooting Guide' published by Jenny Stanford Publishing and distributed by Taylor & Francis. He has a number of research publications in national and international journals. He served petroleum refineries for about 35 years starting from his career up to senior position heading refinery operation, maintenance, safety and training. He also served chemical and petrochemical industries in senior position after a long stint in the refinery sector. He worked as an expert panel member from Centre for High Technologies (CHT) on behalf of Petroleum Ministry for audit of benchmarking performance of Indian refineries.

Acknowledgments

While I started pursuing my profession as a chemical engineer in a petroleum refinery, I eventually got engaged in the production processes of lube base oils. The refinery was a unique refinery in India where all basic process calculations were provided by the licensors of the lube process units which helped me match theories with practices. This has enhanced my core competence in lube base oil processes thereby elongated my tenure in this field for more than a decade during which I had to learn the lube blending process also to ensure making best quality base oils. My career in lube oil production was further strengthened through my association with many international experts in this field in the form of participating and presenting papers in the conferences, seminars, journals, undergoing training in Exxon-Mobil, USA, grass-root lube plant project co-ordination and commissioning, and commissioning of external lube process unit.

It is also observed that most of the refineries are fuel refineries and lube refineries are very few in numbers and capacity as compared to fuel production units; this naturally causes less availability of adequate subject matter expert in this.

Writing this book is an inspiration from above.

My colleagues and seniors always recognized my strength in this field and they extended their co-operation and guidance which also helped to do justice to the book quality. In fact, they inspired me a lot to write this book.

Lastly, it is obvious that without untiring support from my beloved wife, it would have not been possible to come to completion of this book writing. I am sure that readers will be benefited immensely from this book and when only she can be happy.

1 Preliminaries

1.1 DEFINING LUBRICANT AND WAX

Lubricant is defined as a fluid (generally liquid or solid) used in the space between rotary and stationary parts of a machine where, in dynamic condition of the machine, the fluid prevents the friction between two moving parts or between the moving parts and the static parts of the machine by forming a film between these two surfaces thereby saving the life of these machine parts. In this process, the fluid also absorbs the heat generated during the motion of the rotary parts of the machine thereby avoiding the abnormal rise in temperature of the moving parts of the machine. To meet this objective, naturally the fluid should have an adequate viscosity to form the film between the moving parts or between moving and stationary parts of the machine. The lubricant in liquid form is colloquially called lube oil which is generally used in a high-speed machine to facilitate absorption of higher heat generated during high-speed rotation of the moving parts of the machine in addition to lubrication to reduce friction. The lubricant in solid form is called grease and is used in low-speed machine or low-speed components of the machine; here, the grease is a mixture of lube oil and specially prepared soap (acting also as a binder material), and thus concentration of lube oil is less in grease as compared to neat lube oil. There are synthetic solid lubricants also like PTFE, graphite, etc.

The lube oil to maintain its desired viscosity during its life cycle as discussed above should have following additional properties:

- Sustaining the viscosity of the oil in the acceptable range during rise in temperature of the oil caused by heat generation from moving parts of the machine, i.e. the lube oil should have high viscosity index (VI) which represents this property.
- Good non-congealing property, i.e. lower pour point.
- Good thermal conductivity to dissipate the heat generated in the moving parts of the machine.
- Good oxidation stability to avoid degradation of the oil in presence of oxygen.
- Good thermal stability to avoid degradation of the oil with rise in temperature during rotation of the moving parts of the machine.

DOI: 10.1201/9781003334422-1

- Good electrical insulation property while used as transformer oil in the electrical power station/equipment.
- Good cleansing property to wharf out deposits from moving parts of the machine.
- Good corrosion resistance property by forming a protecting film on the surfaces of parts in contact.

The wax, on the other hand, is defined as a solid material which is applied in a non-moving body to provide a very high viscous coating in solid form to protect the inside layer of the main material on which this coating is applied. To have this and other desired properties, the wax should have the following physico-chemical properties:

- Low oil content in the wax to keep it in the solid form in use even with rise in ambient temperature and to keep it hard to avoid loss during handling, i.e. it should have high drop melting point.
- To have no aromatic or any carcinogenic compound in the wax to use it in the food packaging container inside coating.
- Good electrical insulation properties to use it in underground electrical cable coating.

The lubricant and wax are produced mainly from hydrocarbon sources though now-a-days synthetic lubricant and wax have come into play in a very small scale of production and that too in niche sector.

Hence, the chemistry is focused with respect to hydrocarbon sources only and with respect to the physico-chemical properties as discussed above.

1.2 CHEMISTRY

Lubricants and waxes are used in the industries and in the retail sector based on their physico-chemical properties as desired by the respective industries and consumers as discussed above to meet the objectives of application in the respective machine(s) and/or material. Accordingly, the chemistry of the lubricants and waxes are discussed in relation to their physico-chemical properties. The lube oil constitutes mainly base oil(s), and some additives are included to the lube base oil to improve some properties as mentioned above. Lube oils can be of different grades as required for application in different machines. Similarly, the grease also are of different grades in line with their application. These base oils used in lube oils and greases are generally from hydrocarbon sources, and hence, to discuss the chemistry of lubricants (lube oil and grease), it implies that chemistry of base oils should be focused as detailed below:

- **Chemistry of base oils:**
  - Each grade of base oils produced by the lube refineries is sourced from crude oil hydrocarbons, and the base oil contains the following hydrocarbons:
  - n-paraffins having very high VI and excellent oxidation stability but with high pour point.

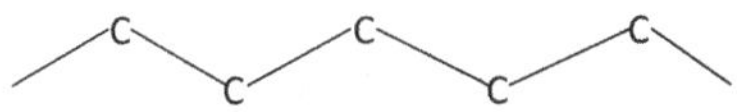

FIGURE 1.1 Molecular structure of n-Paraffin.

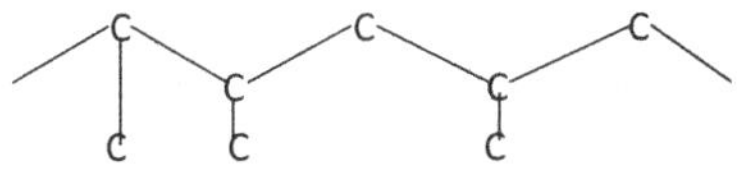

FIGURE 1.2 Molecular structure of iso-Paraffin.

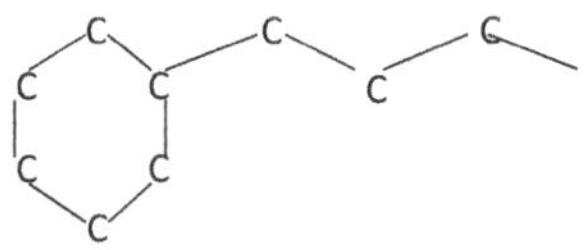

FIGURE 1.3 Molecular structure of Naphthenes.

Note: Demerit of presence of unsaturated side chain(s) in naphthene is that it causes poor oxidation stability to base oil as discussed earlier.

- Iso-paraffins having high VI and excellent oxidation stability as well as low pour point.
- Mono-naphthenes having moderate high VI and excellent oxidation stability as well as low pour point.
- Di/Poly naphthenes having medium low VI, good oxidation stability, and medium pour point.
- Aromatics having low VI, poor oxidation stability but low pour point.

n-paraffins are straight chain saturated hydrocarbon, e.g. methane, ethane, propane, n-butane, and so on. We know methane contains one carbon atom in its molecular formula; similarly, ethane contains two carbon atoms, propane three carbon atoms, and so on. Hence, these compounds are represented as C_1, C_2, C_3, nC_4, nC_5, and so on. The symbolic molecular structure of n-paraffin is shown as Figure 1.1.

Iso-paraffins are saturated hydrocarbons with one or more branched chains starting from Iso-butane, iso-pentane, iso-hexane, and so on. like n-paraffins they are also represented as iC_4, iC_5, and so on. Both n- and i-paraffins in the base oils are found from C_{16} onwards and up to C_{30} in general. The symbolic molecular structure of iso-paraffin is shown as Figure 1.2.

Naphthenes are closed ring saturated hydrocarbons, e.g. cyclo-pentane, cyclo-hexane, and so on. In the base oils, there may be one or two attached naphthene rings that may have straight chain hydrocarbons in their side chain(s). The side chains if unsaturated are not good with respect to oxidation stability of the base oils. The symbolic molecular structure of naphthene with side chain is shown as Figure 1.3.

Aromatics are closed ring unsaturated hydrocarbons with alternate unsaturated bonds inside the closed ring. In the base oils, aromatics have straight chain

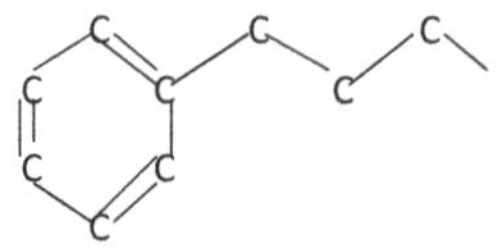

FIGURE 1.4 Molecular structure of Aromatics with side paraffinic chain(s).

Note: Bur if the side chain in Figure 1.4 is unsaturated, it causes poor oxidation stability to base oil; the side chain inside the closed ring as shown in Figure 1.4 being in alternate positions make these bonds stronger in stability thereby not impacting adversely to stability of base oils. The side chain outside the closed ring may have branched chain also as shown in the structure of iso-paraffin earlier.

hydrocarbons in their side chain(s). There may be one or two unsaturated bonds in the side chain but that is undesirable with respect to oxidation stability of the base oils. Sometimes, there may be two aromatic rings attached to each other which is called naphthalene. Compounds with more than two aromatic rings, i.e. anthracene category, do not come in the base oil range and instead this is found in heavy residue after crude distillation. The molecular structure of one aromatic ring with one side chain is given in Figure 1.4.

Like n-Paraffins and iso-Paraffins, side chain(s) attached with a single closed ring (aromatic or naphthene as shown above) contains(s) carbon atoms from 10 to 25 in general.

Sometimes in n-naphthene, in the closed ring, there may be one sulfur atom or nitrogen atom replacing one carbon atom inside the closed ring. For example, with one sulfur atom in a five-membered closed ring of carbon atom, the compound is called thiophane. In six-membered closed rings of naphthene, if one nitrogen atom replaces one carbon atom inside the closed ring, it is called pyridine; if it happens in five-membered rings, it is called pyrrole.

While sulfur contributes favorably with respect to oxidation stability, it may affect adversely in the reducing condition with respect to corrosion by forming hydrogen sulfide. Similarly, nitrogen also affects adversely in reducing atmosphere by forming ammonia. Hence, it is wise to remove both sulfur and nitrogen in base oil making process.

Sulfur and nitrogen may also be present in the side chain(s) of n-paraffins, iso-paraffins, naphthenes and aromatics which are easy to be removed by low severity dehydrogenation reaction unlike in the case of their presence inside the closed ring as mentioned above; in that case, to remove this sulfur and nitrogen, high severity dehydrogenation and denitrification reactions are followed in addition to selecting a specially formed catalyst.

Sometimes, though very uncommon in the case of petroleum crude oil source, there may be one or two unsaturated bonds in the carbon atoms' chain in n-paraffins, iso-paraffins, in side chains of naphthenes and of aromatics; this unsaturation leads to poor oxidation stability of the base oils as already mentioned. However, saturation of these unsaturated bonds is easy to be carried out at moderate severity of hydrogenation reaction process.

- **Chemistry and physico-chemical properties of additives:**
  Various physico-chemical properties required for base oil to have in it as mentioned earlier can't be achieved to full extent through its manufacturing process in a petroleum refinery. Thus, it becomes a challenge to use such base oil in a machine to meet the objective(s). To deal with this problem, many specialty chemicals have been found which on minor quantity addition in single or in combination to the base oil can facilitate achieving the perfect property(s) as required to use the product as a lubricant. These are called lubricant additives. These additives don't react chemically with the base oils but contribute to improvement of above mentioned physico-chemical properties which are additive properties thus to ensure enhancement of the respective properties of the lubricant. In the lubricant(s) formulation(s), choosing right additive(s) is a very important chemistry as well as engineering, and from the subject on 'friction and wear' point of view, 'tribology', covers this aspect in detail with respect to engineering aspect on friction and wear. The present book also includes a chapter on tribology but that only covers a general idea on the subject, because to cover tribology in totality, it is a big subject that could itself be a full volume book.

  To discuss about the chemistry of additives, some representative aspects have been covered in this section as follows.

  Sometimes, base oil so produced in the refinery may need one or two viscosity improvers which reduce the frictional resistance with better co-efficient of friction and also reduces wear in the rotary machine. Molybdenum compounds, viz. molybdenum-dialkyl-thio-phosphate (MoDTP), molybdenum-dialkyl-thio-carbamate (MoDTC), and similar boron derivative have been found best performing in this regard when added to a base oil. Moreover, these compounds have no adverse additive-additive interference as observed.

  The concentration of friction modifiers remains maximum up to 1% in the lubricant; these friction modifiers help in producing multi-grade viscosity lube oils, i.e. both to use the lubricant in summer as well as in the winter. Oxidation stability at high temperature is a very important property as many base oils have been found to undergo oxidation, polymerization, and decomposition at higher temperature in use or in storage.

  A lubricant of multi-viscosity grade which is very common now-a-days invariably contains a viscosity modifying polymer. This polymer helps in minimizing change in viscosity with change in temperature and also changes the rheology of the lubricant with improved wear resistance. The wear resistance depends on the concentration of this polymer in the oil. The decomposition of the polymer at thermal-oxidative conditions, if it happens, would affect the rheology of the oil. The olefin co-polymer used as viscosity modifier additive is commercially known as ethylene-propylene-diene-monomer which has a molecular weight (Mw) of about 260,000, and this is a good quality viscosity modifier (VM). Other viscosity modifiers are polymethacrylates (PMA), olefin co-polymers (OCP), hydrogenated styrene iso-propene (SIP) with [MW of 145,000, 150,000 and 240,000 respectively][13]. The shear stability of the lubricant is influenced by the concentration of VM.

Other additives are: dispersant, detergent, oxidation inhibitor, rust inhibitor, emulsifier / de-emulsifier, antifoamant, etc. Examples of various detergents are: sulfonate detergent, salicylate detergent, phenate detergent, etc. Metal detergents prevent ring sticking under severe high temperature operating conditions; they also reduce carbon and varnish deposits on the engine's pistons.

Rust inhibitors form a surface film to protect the metal of the engine. Examples are: metal sulfonates with high base number, i.e. detergents, ethoxylated alcohol or phenol and alkyl succinic acids and their derivatives.

Emulsifier /de-emulsifiers control the amount of water carried by oil, and maintain liquid-liquid interface.

Antifoamants maintain liquid-gas interface and control foam and air entrainment. Examples are silicone compounds.

Unlike in fuel oil, instead of adding pour point additive in the blend, it is expected that the base oils should have adequate pour point to be used as lubricant except specially required in some cases.

- **Chemistry of wax:**

  Wax is a long chain paraffinic compound or long chain paraffin without or with one or more branch chains or naphthene with side chain(s) of long chain paraffin with or without branch chain with carbon numbers of paraffin chain(s) including branch chain(s) generally remaining at 20 or above which is symbolically expressed as C_{20} or higher. Sometimes, instead of naphthene, there may be one aromatic ring containing the long chain paraffin with or without branch in its side chain.

  If neat concentrations of n-paraffin, iso-paraffin, naphthene and poly-naphthene are concerned, then waxes, produced by solvent dewaxing of very good potential of wax, are found to have the following approximate composition of paraffin wax after hydro finishing of deoiled wax as shown in Figure 1.5.

For micro-crystalline wax (MCW) derived from solvent deoiling of bright stock raffinate, the composition of wax (also called petrolatum[14]) is found to be as shown in Figure 1.6.

In wax from heavier distillates, it is found that percentage of neat 'n' and 'iso' paraffins decreases with increases in mono- and poly-naphthene. With feed source as hydrocracked distillate, concentration of iso-paraffin increases with decrease of n-paraffins and also mono- and poly-naphthene concentration diminishes.

- n-Paraffin 70-75% wt.
- iso-Paraffin 15-20% wt.
- mono-naphthene 5-10% wt.
- poly-naphthene 2-3% wt.

FIGURE 1.5 Composition of paraffin wax.

- n-Paraffin <2% wt.
- iso-Paraffin 24% wt.
- Mono-Naphthene 21% wt.
- Poly-Naphthene 10% wt.
- Aromatics 43% wt.

[14]

FIGURE 1.6 Composition of micro-crystalline wax (MCW).

Note: The aromatic compounds of the product, MCW, can be saturated in subsequent hydro-finishing unit to produce finished on-spec. MCW to achieve good color and oxidation stability in the wax.

Wax is of two categories with respect to its commercial uses, viz. paraffin wax and micro-crystalline wax (MCW).

When it is only single chain paraffin without or with minimum branch chain or without any naphthene or aromatic as a part of the paraffin chain, the wax is called paraffin wax, and with respect to crystallography of the wax, the crystal structures are macro in sizes.

On the contrary, when there are more branch chains in the base paraffin compound or there are more naphthene rings with side chain long paraffin, the wax crystallography changes to micro-crystalline in nature, and hence, it is then called MCW.

The molecular structures of paraffin wax and MCW are same as shown earlier in the sub-Section 1.2 except the carbon of the paraffin chain with or without branch chain should be C_{20} or more as mentioned above also.

Regarding physico-chemical properties of paraffin wax and MCW, it can be said that paraffin wax is hard but brittle with drop melting point as allowed for commercial uses is minimum 45°C depending upon ambient temperature while MCW is soft but not brittle with higher drop melting point at and above 70°C. Due to these differences in the properties, the paraffin wax and MCW have different application in the industries and retail sector.

In paraffin wax produced synthetically, the wax may be softer if in the polymerization process of n-paraffin, the polymerization stops prematurely before desired completion of polymerization making the wax so produced softer due to low molecular weight or polymerization process generates some branch chains, the wax so produced would be micro-crystalline in nature resulting in the production of MCW.

1.3 TECHNOLOGIES

Most of the lubricant/lube base oil and wax production are done with feedstock as derivatives obtained from mineral oil, i.e. petroleum crude oil while recently these are also being produced from vegetable oil as feedstock but the market contribution from this vegetable oil sector is negligible as compared to petroleum crude oil source. However, the base oil and wax produced from vegetable feedstock are biodegradable and hence, environment friendly. A brief on the technologies followed with petroleum feedstock are described below while that with vegetable feedstock is discussed only in the respective chapter as the same has just started to be commercially established in a large scale unlike in petroleum sector.

- **Lube base oil production:**

  The boiling range of lube base oils are generally in the ASTM boiling range of 370°C as IBP (initial boiling point) up to 500°C as FBP (final boiling point), the vacuum distillate(s) of petroleum crude oil are used as the feedstock in the manufacturing process of lube base oil. However, vacuum distillate(s) from some petroleum crudes, though meeting the above boiling range, still are not suitable to use these in the lube base oil production because these vacuum distillate(s) don't have the desired physico-chemical properties like viscosity and minimum viscosity index (VI) to process it for further improvement of VI.

  The conventional manufacturing processes being followed since about last seven decades are:

  - Aromatic extraction from the vacuum distillate(s) in order to remove the aromatics which though having higher viscosity, have very poor VI. Common solvent used in the extraction processes are either Furfural or NMP (n-methyl pyrolidone). The oil so produced after aromatic extraction is called raffinate. The commercial technologies in detail have been covered in Chapter 2.
  - The raffinate produced from aromatic extraction unit is further processed in solvent-dewaxing unit to remove a good amount of waxy paraffins in order to improve the congealing property of the oil, i.e. to get the final oil with lower pour point to achieve the fluidity of the lube base oil. Though paraffins have very good VI (but with a bit lower viscosity), this VI sacrifice is followed due to reason as mentioned above. The oil so produced is called dewaxed oil (DWO). The commercial technologies in detail have been covered in Chapter 2.
  - The DWO produced from the dewaxing unit is found to have some dark color or poor color stability and not good for use in the long run in the machine. Also, sometimes it generates some gum like compound on long storage or on use for long. The former is due to presence of nitrogenous compound in the DWO and the latter is due to presence of some olefin compounds, i.e. with unsaturation in the paraffin chain in the lube base oil. Also, a small percentage of sulfur (1 to 2%) in the lube base oils, but if it is more than this, the same needs to be reduced as though sulfur contributes to good oxidation stability to the base oil, it has negative contribution by causing corrosion in the long run because due to thermal degradation in reducing atmosphere in the long run, hydrogen sulfide may get produced which in turn causes corrosion to the machine.
  - The above-mentioned sulfur and nitrogen reduction and olefin saturation are achieved by processing the DWO in a mild hydro-treatment unit called

hydro-finishing unit where at a certain pressure and temperature hydrogen is used to react with sulfur, nitrogen and olefin bonds to get the desired product. The unit is usually called a hydro-finishing unit. The commercial technologies in detail have been covered in Chapter 2.

Alternative to the above dewaxing processing of raffinate, now-a-days catalytic processes have become commercially popular which isomerizes these waxy n-paraffins thus reducing pour point of the lube base oil to a great extent though viscosity index (VI) of the lube base oils gets marginally reduced and viscosity marginally increases due to branching of n-paraffins. Also, in this process, while the by-product, oily wax is not produced thereby contributing to increase the yield of lube base oil, the ultimate yield of the main base oil does not increase due to formation of other by-products like lower viscosity base oil as an additional product plus some diesel, naphtha and LPG caused by cracking process as inevitable in the technology. This technology can be used to make lube base oil from vacuum distillate(s) having no lube potential as discussed earlier through its processing first in hydro-cracking unit (meant for production of fuels like diesel mainly and naphtha, LPG in addition with waxy residue either to further hydrocrack or take it out for processing in catalytic dewaxing unit as mentioned above), and the residue obtained from there is processed in catalytic dewaxing unit as mentioned above. The catalytic dewaxing unit is popularly called a catalytic iso-dewaxing unit (CIDW). The commercial technologies in detail have been covered in Chapter 2.

In addition to above, there is standalone lube base oil produced synthetically through stoichiometric reaction process between two specialty hydrocarbons; example is: PAO (poly alpha olefin) which is also called Group-IV base oil or lube oil as discussed in Chapter 2. Also, there are other synthetic lube oils like different esters, polyols, etc. which are called Group-V lube oils or base oils as also discussed in Chapter 2. Bio-lubricant produced from vegetable oils also fall under Group-V lube oil though these can't be called synthetic lube oils because these are not produced through chemical reactions between two or more hydrocarbons and instead these are produced through physical mixing and purification processes with vegetable oils used as feedstock.

- **Additive production:**

  To get the final lube oil / lubricant, specific additive(s) is/are to be added to the base oils except Group-IV base oil (PAO) which has poor solubility to additive; hence PAO is to be used as a standalone lubricant to the machine as designed to use the same. But Group-V base oils have good solubility with additive(s) and hence, their properties can be modulated to the requirement of wide range of machines except bio-lubricants which have their inherent limitation w.r.t. their solubility in additive(s) and hence find limited application.

  Regarding production technologies for additives, there is no common technology unlike for Group-I, II and III base oil production; rather, the additives are produced through stoichiometric processes, i.e. by chemical reactions between two or more chemicals at specific temperature, pressure, and residence time of reaction in a specially designed reaction vessel along with its auxiliaries; the reaction processes are developed based on research and development in 'tribology' as discussed further in Chapter 6 while blending of these additives with base oils to produce final lubricants are discussed in Chapter 3.

- **Wax production:**
  Waxy hydrocarbons have also similar ASTM boiling range like lube base as discussed in sub-Section 1.3 above. The vacuum distillate(s) of petroleum crude oil are used as the feedstock in the manufacturing process of wax. However, vacuum distillate(s) from some petroleum crudes, though meeting the above boiling range, still are not suitable to use these in the wax production because these vacuum distillate(s) don't have adequate wax potential as well as they don't have the desired physico-chemical properties like drop melting point, ductility, minimum residual oil content achievable while being produced through separation of inherent oil content in the feedstock.

  If the vacuum distillate has very high aromatic content, then aromatic extraction to be followed for removal of aromatics in order to achieve desired color (yellow or brighter) as described in sub-Section 1.3 above followed by solvent deoiling of wax (synonymous to solvent dewaxing of raffinate as discussed in sub-Section 1.3 above).

  If the vacuum distillate has very low concentration of aromatics and asphaltene, then the aromatic extraction process as mentioned above can be bypassed and the vacuum distillate can be directly processed in solvent deoiling. The technology of solvent deoiling is same as that of solvent dewaxing as mentioned in the sub Section 1.3 above except that unlike in solvent dewaxing where dual solvent like MEK (methyl ethyl ketone) and toluene are used, here a single solvent like MIBK (methyl iso-butyl ketone) is used and the solvent distribution ratios at different locations are different than those in solvent dewaxing, and also, crystallization and filtration temperatures are kept higher than those in solvent dewaxing. Hence, as such, technology wise, solvent dewaxing and solvent deoiling are same except in operating parameters which are discussed in detail in Chapter 4.

2 Lube Oil Base Stock (LOBS)

2.1 PHYSICO-CHEMICAL PROPERTIES OF LOBS AND SIGNIFICANCE

The various important physico-chemical properties of LOBS which play the roles in their application to the machines are discussed as below.

- **Viscosity:**

LOBS which are the main constituent of lube (a colloquial term of lube oil) finds its application meaningful due to its viscosity property mainly; viscous liquid only can create a retaining film of liquid on the surfaces of the stationary and moving parts in contact in a machine thereby avoiding metal to metal contacts between relatively moving parts, and also in this process it absorbs the heat generated due to kinetic motion of the lube and moving parts of the machine. The heat absorbed in the lubes are subsequently dissipated by establishing a circulation system of the lubes where there would be a cooling unit thus getting the cooled lubes back to the machine. For smaller machines, generally such a circulation system is not required for the lubes; the non-insulated housing of the lubes with an open vent facility becomes enough to dissipate the mild heat generated in friction.

As one can understand depending upon the relative speed of the machine parts and their resultant thrust on the rotation/movement, viscosity requirement for the LOBS would vary to ensure proper retention of the liquid film on the parts in contact. Due to this, there are various grades of LOBS with respect to viscosity, say, low viscosity grade, medium viscosity grade, and high viscosity grade. Also, it is true that viscosity changes with change in temperature; hence, while defining viscosity requirement, the value of viscosity is given at a particular reference temperature.

Generally, there are mainly three viscosity grades, viz. 150N, 500N and 150 BS though there are other viscosity grades like 100N, 800N and 1300N in the industry. Here, 100N is lighted LOBS in respect to viscosity, then come 150N, 500N, 800N and 1300N whereas 150BS grade falls under heavy lubes category with respect to viscosity.

DOI: 10.1201/9781003334422-2

TABLE 2.1
Viscosity, cst. at 98.9°C vs. Title of LOBS

Grade	Viscosity in Centistokes at 98.9°C.
100N	2.5–3.5
150N	4.5–5.5
500N	10–11
800N	12–13
1300N	14–15
150BS	29–32

Here, the terminology, N, denotes the LOBS is produced from vacuum distillate source in crude oil refinery, and in the processes, concentration of aromatic compounds (where polarity is high) got reduced to approach neutrality in polarity.

Viscosities of the above grade are set internationally as shown in Table 2.1.

Among the hydrocarbons, viscosity value changes in the increasing order from olefinic, paraffinic, iso-paraffinic, naphthenic, aromatic, and asphaltic compounds. Similarly, in the same category of hydrocarbon group, viscosity increases with increase in molecular weight. Lubricant is a mixture of above groups of hydrocarbons mainly having paraffin, iso-paraffin, naphthenic, and aromatic compound with paraffinic, iso-paraffinic, and naphthenic compounds in the carbon numbers range of C_{16} to C_{22}, aromatic compound of C_6 structure with paraffinic, iso-paraffinic and/or naphthenic side chains which provide the desired viscosity in the lubricant.

Viscosity is measured in different scales, viz. Centipoises or Centistokes, Saybolt Seconds Universal (SSU or SUS), Degree Engler (°E), and Redwood seconds. Centipoise divided by density in gm/cc is called Centistoke.

- **Viscosity index (VI):**

Viscosity of a liquid generally changes with change in temperature, i.e. when temperature increases, viscosity decreases, and vice versa. Hence, the lubricant used in a machine as mentioned above will undergo a reduction in viscosity due to increase in temperature caused by kinetic motion of the lubricant and moving parts of the machine in contact thus making the whole purpose of lubrication ineffective. So, one should select the lubricant such that there is minimum change in viscosity with change in temperature, and thus viscosity index (VI), vital property of lubricant, has come into place which is defined as: rate of change in viscosity of the lubricant with change in temperature in use or operation; in this correlation, change in viscosity low means, value of VI would be high and vice versa. VI is a dimensionless value of a lubricant or LOBS. The scale of the values of VI can be between 0 to >100 depending upon the liquid in use. For lubricant used in the market, it generally remains within 90 to 120. The correlations commonly available for VI are as follows:

for oil of VI between 0 to 100:

$$VI= ((LV - UV) / (LV - HV)) \times 100 \quad (2.1)$$ [1]

for oil of VI above 100:

$$\text{Log } UV= \text{Log } HV - N \text{ Log } YV \quad (2.2)$$ [1]

where,

LV is the kinematic viscosity at centistokes (cst.) at 40°C of a reference oil with 0 VI which has same value with oil having viscosity at 100°C for which VI is to be calculated above.

HV is the kinematic viscosity at centistokes (cst.) at 40°C of a reference oil with 100 VI which has same value with oil having viscosity at 100°C for which VI is to be calculated above.

UV is the kinematic viscosity at centistokes (cst.) at 40°C of the oil for which VI is to be calculated above.

YV is the kinematic viscosity at centistokes (cst.) at 100°C of the oil for which VI is to be calculated above.

$$N = \text{Log } (0.00715 (VI - 100) + 1) \quad (2.3)$$ [1]

The equations are cumbersome to follow in regular use. In practice, nomograph (or also called a nomogram) is prepared for a category of a liquid, say, for Newtonian fluid having similar physico-chemical properties, say, for hydrocarbon oil derived from petroleum crude, or say for a pure component liquid and available in literature for use. In the nomograph, generally, three vertical scales of measurement are there with one scale for viscosity at 40°C, second one is viscosity at 100°C and third one is VI of the liquid. So, knowing any two values, the value of the third one can be obtained by drawing an intersecting straight line passing through all these three vertical scales of measurement.

Among the hydrocarbons, VI changes in the decreasing order from olefinic, paraffinic, naphthenic, aromatic, and asphaltic compounds. But in Olefinic compound, oxidation and thermal stability are less.

So, in producing/using lubricant, focus is given to have maximum concentration of paraffinic or iso-paraffinic compound in the carbon number range of C_{16} to C_{22} to get desired viscosity and VI simultaneously with moderate concentration of naphthenic compounds in the same carbon number range but keeping minimum concentration of aromatic compounds to offset its negative impact in VI.

- **Pour point:**
  It is understood from above that lubricant contains major quantity of paraffin of carbon number range of C_{16} to C_{32} whether as standalone or as a side chain of naphthene. But paraffins in this range are congealing in nature thereby preventing fluidity of lubricant and therefore defeating the whole purpose of

lubrication. Pour point is a temperature at which the liquid would cease to flow normally and this temperature is just a few degrees above the freezing point (or say congealing point) where the liquid gets solidified; this happens at lower temperature. In a cold climate or in winter season, the ambient temperature falls down to sub-zero level thus the lubricant in the machine in idle condition in there would freeze preventing the start of the machine successfully. Pour point of lubricant varies from (–)3°C to (–)12°C and sometimes to (–)50°C depending on types of the machine and ambient condition. Very commonly, most of the lubricant or lube oil base stock (LOBS) have a pour point of (–6) °C.

From the above, it is evident that in the LOBS and lubricant, paraffin content should be reduced by compensating with iso-paraffins and with moderate naphthene and minimum aromatic to optimize all the above three properties to meet the specifications of lube oil.

- **CCR (carbon conradson residue):**

We know aromatic compound contains relatively higher numbers of carbon with respect to hydrogen as compared to paraffin and iso-paraffin; also, aromatic has higher carbon content as compared to naphthene; similarly, asphaltene contains higher carbon content as compared to aromatic. Lubricant in use gets heated in use as discussed earlier; so, there would be tendency of thermal cracking of the lubricant in operation; so, with more carbon content in the lubricant, there would be more thermal cracking resulting in carbon deposit, i.e. charring in the lubricant thereby affecting the properties of the lubricant; also, there is a chance of carbon deposit in the machine parts in contact. Due to this, the aromatic and asphaltene content should be as minimum as possible in the lubricant. CCR is defined as carbon residue remains after carrying out thermal cracking of the lubricant/LOBS sample in specially designed apparatus as defined in the ASTM standard. It is expressed in %wt.

For good lubricant/LOBS, the value of CCR varies between 0.03%wt. up to 0.3%wt. starting from light grade to heavy grade of lubricant/LOBS.

- **Color:**

Though color can't be a specific requirement for the lube oil in a true sense, it is true that color is contributed by the nitrogenous compound mainly with sulfur compounds contributing to a mild greenish color. The nitrogen content beyond a limit is harmful to lubricant as the lubricant in the long run of use would produce some nitrogenous compounds which would damage the machine parts in contact.

Color of the LOBS/lubricant is generally measured in line with ASTM standard and the value varies from 0.5 ASTM up to 5 ASTM scale in the order of light lube to heavy lube as well as depending upon whether Group-I or Group-II LOBS is produced. Color value of Group-II LOBS is always lower than that of Group-I LOBS as in the process of Group-II LOBS production, hydro-treating of the feedstock is carried out to saturate the aromatic to naphthene to increase the value of VI while in that process simultaneously nitrogen gets converted into ammonia which in turn gets removed from the separating column/vessels in the process.

- **Color stability:**
  Sometimes initial value of color in the lubricant/LOBS is low, i.e. the color is very light yellow, but after use for a few days, the color darkens and that is not acceptable as discussed above. Hence, the property specification like 'color stability' has come into place which is determined by heating the sample at 100°C for 24 hours in a standard apparatus and measuring the color thereafter. It is also expressed in the scale as discussed above.
- **Oxidation stability:**
  Though sulfur contributes to a very good property, viz. oxidation stability, which is required in lubricant/LOBS, higher sulfur content has however a negative effect like its slow decomposition to hydrogen sulfide and in the long run leads to corrosion by reaction with iron which is present in machine parts during operation and heat generation. Aromatics and unsaturated compounds like olefins, if any, and/or unsaturated naphthene, have a tendency to oxidize in presence of air in the long run and will form epoxides which in turn will polymerize into gum and ultimately would precipitate in the storage tanks or inside the idle machines in the long run thereby damaging the main properties of lubricant/LOBS, i.e. lube containing such components have poor oxidation stability. Sulfur having surplus electrons in their molecular orbitals can share these to aerial oxygen thereby preventing the oxygen from attacking the unsaturated compounds present in the lubricant. In absence of adequate sulfur concentration in the lube oil, some external anti-oxidant chemical is to be dosed into lubricant or finally produced LOBS to ensure the role of sulfur as explained above. Commonly marketed anti-oxidant into this industry is DBPC (di-butyl para-cresol). Hence, either sulfur content should be optimized or if it had to be reduced to a very low level due to the process followed for LOBS production, then external dosing of some oxidation stabilizer as mentioned above is to be dosed to the LOBS to enhance oxidation stability property of the LOBS/lubricant.
- **Flash point:**
  Flash point is a temperature generally measured in centigrade scale as per ASTM standard and there are two methods of measurement in ASTM standard, viz. PMCC method and COC method. We know lower flash point is always unsafe with respect to possibility of explosion in undesired situation like getting a visible or invisible ignition spark in the vicinity. For transport fuel, the value limits are kept lower as the liquids are lighter in those case which obviously have lower flash point and thus accordingly taken care in the automobiles. But the lubricant is not used as burning fuel rather used as a stable non-fuel liquid which should remain in position without damaging it's properties, i.e. its flash point also should be much higher to avoid any hazardous explosion due to it getting heated in use. Lubricant/LOBS being of higher molecular weight as discussed earlier would have higher flash point naturally but in the process of manufacturing LOBS, there may be formation of some lighter hydrocarbon reducing only its flash point. Hence, process must ensure removal of such lighter hydrocarbon before concluding the process. Accordingly, flash point

values have been specified in market specification of LOBS and Lubricant. The value varies from about 150°C up to 300°C depending upon lighter to heavier grades.

- **Other properties:**

Though the above properties are the main properties of LOBS/lubricant, there are other properties of the LOBS and lubricant in the standard specification under different grade names as discussed earlier. These are as follows:

Saturate content, %wt. min.
Sulfur content, ppm or %wt. max.
Ash content, %wt. max.
TAN (total acid number), mg KOH/gm sample, max.
Inorganic acidity, mg KOH/gm sample, max.
Water content, ppm, max.
ARV (air release value) at 50°C, max.
Nitrogen content, ppm, max.
Emulsion test @54°C, max.
Emulsion test @82°C, max.
NOAK, %wt., max.
Volatility @15% loss, °C, min.

These properties are not a serious concern as these are generally achieved in the process of manufacturing of LOBS and not monitored on a daily basis and instead measured after full tank production is done and before selling/dispatch.

The impacts of composition of LOBS/lubes to various physico-chemical properties are explained in Figure 2.1.

2.2 CLASSIFICATIONS AND CHEMICAL COMPOSITION OF LOBS

Different grades of LOBS used in making different grades of lubricant as required for different categories of machines and automobile engines as follows.

These different grades are characterized by their physico-chemical properties as established to be required for different categories of machines and automobiles. Broadly, these grades are determined by the categories of automobiles like light motor vehicle (LMV), medium motor vehicles (MMV) and heavy motor vehicle (HMV) and similarly in the case of industrial machines and turbines. Based on these requirements, it is found that light to heavy lubricants, viz. lower to higher densities and viscosities of lubricants/LOBS are required respectively. But this is not all; in the above categories of machines, it is also found that there are other properties of the lubricant/LOBS, viz. VI, pour point (mainly) are dictated by the specific categories of machines. Accordingly, LOBS are classified in different grades as per these requirements. Also, in the industry, it is found that in most of the cases, color, color stability and oxidation stability properties are must to be followed. We also have understood that these compounds are present mainly in the side chains of aromatic compounds and inside the structural ring itself in heterocyclic compounds. These

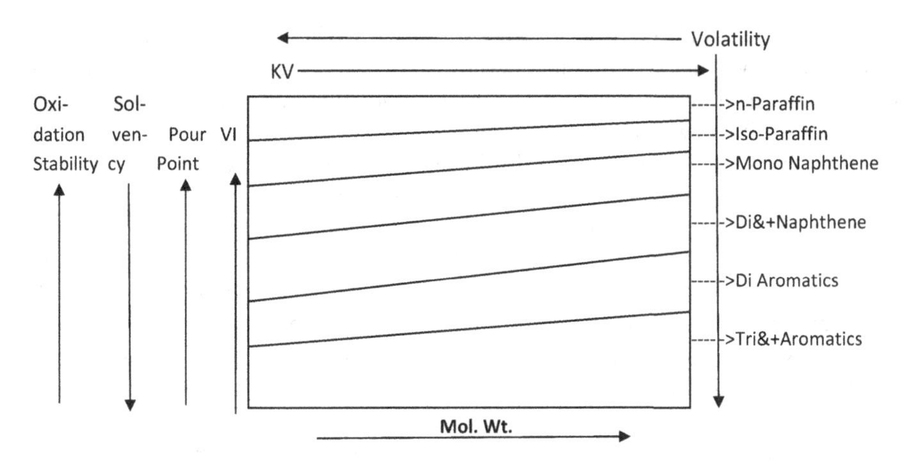

FIGURE 2.1 Relations between compositions and properties of LOBS.

Note: the saturate content and sulfur content are very important properties for Group-II and Group-III base oils but not that much stringent for Group-I LOBS. The importance of these two properties have been understood earlier, also these two properties have been discussed separately in the manufacturing process of Group-II and Group-III LOBS.

Source: [2].

nitrogen compounds contribute to color and color stability problem. Accordingly, in LOBS production, to prevent oxidation of the lube oil in the long run, unsaturates in the form of straight chain hydrocarbons, inside heterocyclic compound or the side chains of aromatic content should be minimum. Also, unsaturate means there is electrical polarity in the compound, so removal of such compounds means polarity is reduced in the lubricant/LOBS. That is why, LOBS produced after processing the feedstock in the intermediate process unit, viz. aromatic extraction unit where some solvent is used to remove the aromatics and unsaturates, is called solvent neutral LOBS. If instead of processing through aromatic extraction unit, the feedstock is processed in catalytic hydro-treating unit using hydrogen to saturate the unsaturates and aromatics, the LOBS produced still can be called neutral LOBS but not solvent neutral. So, in a nut shell, physical separation process yields solvent neutral LOBS whereas the reaction process yields neutral LOBS.

Based on the above discussions, different grades of LOBS are classified as: 70(SN) or 70N, 100(SN) or 100N, 150(SN) or 150N, 500(SN) or 500N, and so on. However, in producing heavy grade of LOBS, processing the feedstock in aromatic extraction is a must in the production process; the solvent extracted LOBS intermediate produced is processed in solvent dewaxing unit or processed in catalytic hydro-dewaxing unit to improve pour point property of LOBS, the grade name of LOBS does not change, for example, 150BS which signifies heavy grade of lube having kinematic viscosity of about 28.5 to 32 cst. (centistokes) @98.9°C or @210°F. The VI requirement of this grade is 95 which can be produced by processing de-asphalted oil (DAO) in aromatic extraction unit (AEU) followed by solvent dewaxing unit

(SDU) or instead of processing through solvent dewaxing unit, processing in catalytic hydro-dewaxing (CIDW) unit where it also can't increase VI more than 95 except yield pattern in catalytic dewaxing unit would be different, i.e. higher than that from solvent dewaxing unit.

Here, feedstock means vacuum distillates, i.e. distillates obtained from vacuum distillation unit, of different viscosities range. For 150BS production, vacuum residue is processed in propane de-asphalting unit using propane as solvent and product obtained called de-asphalted oil (DAO) is then subsequently processed in AEU followed by processing in SDU or CIDW. The nomenclature terminology like 70, 100, 150, 500, and so on which are followed by the abbreviated alphabets (N or SN) represent the viscosities of the respective LOBS in Saybolt Universal seconds (SUS) scale which is near to the respective values of kinematic viscosities measured at 40°C. However, in their property tables, kinematic viscosities, cst. @40°C are also provided. Here, N and SN represent neutral or solvent neutral as explained above. In the nomenclature of 150BS, the term represents 'Bright Stock', a heavy category of base oil having kinematic viscosity of about 30 to 32 cst. @210°F or about 900 cst. @40°C (i.e. @100°F). This 900 cst. is equivalent to the viscosity value in °E in Degree Engler scale, which is about 150.

- **Classification of LOBS:**

Based on following AEU (aromatic extraction unit) and/or SDU (solvent dewaxing unit) processes, i.e. physical separation processes, or reaction process like CIDW (catalytic iso-dewaxing) as discussed above, the LOBS produced would have different physico-chemical properties mainly with respect to concentration of saturate content and sulfur content in LOBS with value of VI in LOBSs different in each case. Accordingly, LOBS is also classified by API (American Petroleum Institute) in terms of saturate, sulfur and VI values and are shown in Table 2.2, except Group-V LOBS which is generally synthetic lubricant produced using some specific pure component compounds having superior lubricating properties than all other grades.

- **Chemical composition of LOBS:**

Chemical composition of various classes/groups of LOBS are shown in Table 2.3.

Instead of the word, 'Group', the abbreviated words, 'API' is also used in the nomenclature of classification above as the classification was introduced by American Petroleum Institute (API).

Group-I category of LOBSs are produced following processing vacuum distillate in ARU and SDU as explained above. However, base produced from SDU is to be further treated in mild hydro-treating unit (HFU) to improve the color and oxidation stability properties as discussed earlier.

Group-II and Group-III LOBSs are produced following processing of vacuum distillate with or without processing in ARU followed by processing in CIDW. The feedstock qualities and operating conditions in CIDW will be different in producing Group-II and Group-III LOBS. Group-II and Group-III LOBSs can also be produced

TABLE 2.2
API Classifications of LOBS

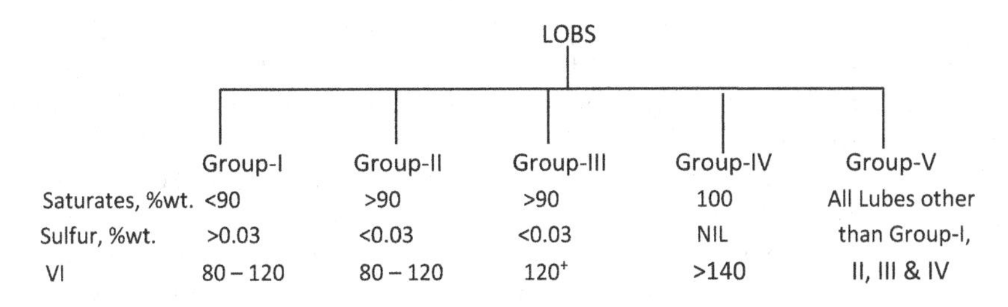

LOBS

	Group-I	Group-II	Group-III	Group-IV	Group-V
Saturates, %wt.	<90	>90	>90	100	All Lubes other
Sulfur, %wt.	>0.03	<0.03	<0.03	NIL	than Group-I,
VI	80 – 120	80 – 120	120^+	>140	II, III & IV

Note:

Both Group-IV and V LOBSs are called synthetic lubricant.

Both Group-IV and V LOBSs can be used as such as lubricant for end use in engines; also, these can be used in blending with base oils from other groups as mentioned above.

Pour point of <(–)57°C., Volatility of <6.5%wt and very high oxidation stability of more than 13000 hrs. can be achieved with Group-IV LOBS.

Group-V LOBS have no specification unlike other groups as mentioned above but generally they exhibit similar properties with respect to viscosity and VI, and pour point like in Group-IV, but additionally, they have better properties with respect to biodegradability better than all these four groups.

TABLE 2.3
Composition of Various API Groups of LOBS

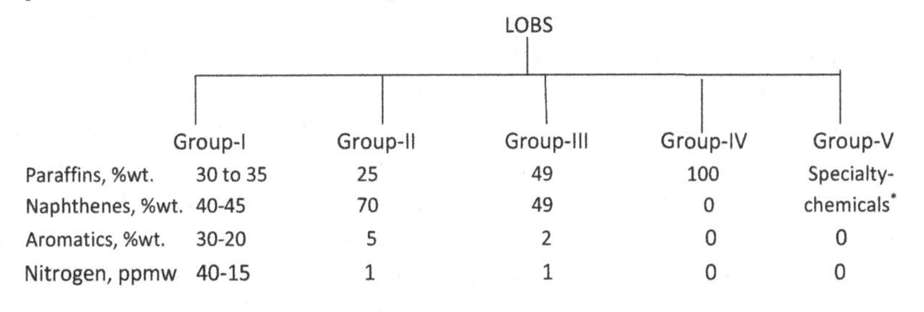

LOBS

	Group-I	Group-II	Group-III	Group-IV	Group-V
Paraffins, %wt.	30 to 35	25	49	100	Specialty-
Naphthenes, %wt.	40-45	70	49	0	chemicals*
Aromatics, %wt.	30-20	5	2	0	0
Nitrogen, ppmw	40-15	1	1	0	0

*It can be Bio-Lubricant also.

Source: [3].

by processing even non-lube vacuum distillates in lube hydrocracker and the residue of the hydrocracker, i.e. hydrocracker bottom can be further processed in SDU only to remove excessive wax in the hydrocracker bottom. These various processes have been separately discussed in respective sub-chapters.

Group-IV and V LOBS also called synthetic lubes are made up of a pure component except in Group-V where vegetable oils also can be used as feedstock. In Group-IV and V lubes can be used as a blend with other base oils.

Example of Group-IV LOBS is 'poly alpha olefin' (PAO).

PAO has excellent lubricating property, like kinematic viscosity of about 14 cst. @40°C with a very high VI >140 and flash point at about 230°C, and can be used in high end automotives requiring special lubricant. PAO is also used in the blend with Group-II and III LOBSs to produce superior lubricant.

Examples of Group-V Lubes are: synthetic esters like, polyol like PAG (poly alkylate glycol), phthalate, etc. and sometimes vegetable oils.

Group-V Lubes may or may not have lower pour point than other groups, but its anti-friction property, i.e. friction co-efficient is better than that of Group-IV lubes. Group-V lubricant has the best biodegradability as compared to all other groups. Vegetable oils also fall into Group-V category if these can be used as lubricant improving their other properties like pour point, neutralization numbers, etc.

2.3 STANDARDS / MARKET SPECIFICATIONS OF LOBS

As discussed earlier there are two types of LOBSs, one category is solvent neutral (SN) and other category is neutral (N) however there is a third category produced synthetically from one or two pure components which are called special grade of LOBS like Group-IV and Group-V LOBS as discussed earlier. The SN refers to virgin LOBS, i.e. produced by physical separation of aromatics by use of some solvent, and aromatics means more polarity due to presence of unsaturated bonds in their chemical structure, the LOBS produced in this process is called solvent neutral (SN). But if the aromatics are saturated by chemical process like hydrogenation, the LOBS produced is called only neutral (N).

There is no EURO or BIS specification for LOBS except API classification as shown earlier in Table 2.3. But there are specifications of finished lubricants for end use in various automotives, industrial engines, and turbines. Also, many refineries have their own lube oil blending plants to produce these finished lubricants by using these LOBSs; hence, if LOBSs produced by their refineries fall short of quality deviations to minor extent, they can manage the same in their finished lube production process by adding some additives while carrying out blending process. However, the basic ingredient for the process, i.e. LOBSs, should be within some range with respect to some particular properties so that these LOBSs can be used as feedstock to the blending plant. Similarly, if the refineries have to market their base oils to some external lube blending or marketing companies, they also have to maintain the base qualities in this direction. With this background typical marketing specifications of LOBSs have been observed to be applicable in the industries.

The typical specifications of N category of LOBSs produced by catalytic dewaxing process as dominant in the marketing are given in Table 2.4.

TABLE 2.4
Properties of Various Grades of Hydrotreated LOBS

Attribute	100N	150N	500N	150BS
Appearance @20°C	Clear and Bright	Clear and Bright	Clear and Bright	Clear and Bright
API Gravity	To report	To report	To report	To report
ASTM Color, max.	Water white	Water white	Water white	<4.5
Color Stability for 24 hours @100°C, ASTM, max.	0.5	1	1	<5.5
Viscosity, cst. @40°C	19–21	28.8–32	95–105	To report
Viscosity, cst. @100°C	To report	To report	To report	29–32
VI, min.	110	105	105	95
Flash Point, COC, °C, min.	180	205	232	280
Pour Point, °C, max.	(–)15	(–)12	(–)9	(–)6
TAN, max.	0.03	0.03	0.03	0.03
Inorganic Acidity, mg KOH/gm	Nil	Nil	Nil	Nil
Air release value (ARV) @50°C, minute, max.	1	1	3	---
Copper strip corrosion @100°C For 3 hours, max	1	1	1	1
Ash content, %wt., max.	0.003	0.003	0.005	0.005
CCR, %wt., max.	0.03	0.05	0.1	0.2
Water content, %volume or wt.	Nil	Nil	Nil	Nil
Sulfur, %wt., max. °	0.01	0.01	0.01	<0.6
Nitrogen, ppm, max.	15	15	15	To report
Emulsion test @54°C, max.	5	10	---	---
Emulsion test @82°C, max.	---	---	5	15
Aromatics (CA), %wt., max.	0.2	0.2	0.2	To report
NOACK, %wt., max.	15	15	15	To report
Volatility @15% loss, °C, min.	370	380	435	---

Similarly, the typical specifications of SN category of LOBSs observed to be followed in the industries are shown in Table 2.5.

2.4 MANUFACTURING TECHNOLOGIES FOR LOBS

2.4.1 Processes Followed for Group-I LOBS Production

Group-I LOBSs are produced through physical separation processes/technologies using vacuum distillates as feedstocks. Various distillate cuts/products are obtained by vacuum distillation of residue (also called long residue) of crude distillation unit. With specific origin of crude source used, it is observed that these distillate cuts are viscous in nature which can be simply understood by taking the samples in hand which are found sticky in finger forming oil film in the finger and also that even after

TABLE 2.5
Properties of Various Grades of Solvent Neutral LOBS

Attribute	100SN	150SN	500SN	150BS
Appearance @20°C	Clear and Bright	Clear and Bright	Clear and Bright	Clear and Bright
API Gravity	To report	To report	To report	To report
ASTM Color, max.	1.0	1.5	2.5	<4.5
Color Stability for 24 hr @100°C, ASTM, max	1.5	2.0	3.0	<5.5
Viscosity, cst. @40°C	To report	24–32	90–100	To report
Viscosity, cst. @100°C, min.	4.0–4.5	5.0–7.0	10–11	29–32
VI, min.	95–100	100–110	97	95
Flash Point, COC, °C, min.	190	210	232	280
Pour Point, °C, max.	(–)12	(–)9	(–)6	(–)3
TAN, max.	0.05	0.05	0.05	0.05
Inorganic Acidity, mg KOH/gm	Nil	Nil	Nil	Nil
Copper strip corrosion @100°C For 3 hours, max	1A	1A	1A	1A
Ash content, %wt., max.	0.01	0.01	0.01	0.01
CCR, %wt., max.	0.01	0.03	0.08	0.3
Water content, %volume or wt.	Nil	Nil	Nil	Nil
Sulfur, %wt., max.	0.5	1.0	1.5	2.0
Nitrogen, ppm, max.	50	100	150	To report
Emulsion test @54°C, max.	5	10	---	---
Emulsion test @82°C, max.	---	---	5	15

warming that sample this filming nature persists. The various cuts that are obtained vary from light viscosity to heavy viscosity thereby named as very light viscosity oil, for example, spindle oil (SO), light viscosity vacuum oil (LO), intermediate viscosity oil (IO), and heavy viscosity oil (HO). Also, the vacuum residue obtained from vacuum distillation also found to contain some heavier viscous oil which could not be separated through distillation. Accordingly, applying liquid-liquid extraction of this vacuum residue using liquid propane as solvent, this heavier oil also could be separated; the product oil from this process is called de-asphalted oil (DAO), and the unit is called propane de-asphalting unit (PDAU). Crude source quality and operating conditions controlling ASTM boiling range of vacuum distillation unit (VDU) decide the viscosity of this oil mainly.

We understood earlier that viscosity is not the only property of LOBS, so many other properties are there in the specifications given earlier out of which some important properties are VI, CCR, pour point, color, color stability, and oxidation stability whose importance have also been discussed. Hence, subsequent physical separation processes have also been established by the technology houses and applied in the industry to improve these important properties of SO, LO, IO, HO, and DAO. Unless crude source is not changed, then achieving these properties, viz. VI, CCR, pour point, color, color stability, and oxidation stability would fix other properties as shown in the specification table if operating conditions of the decided process units are not deviated from normal conditions set by the technologies and experiences.

First process applied to these distillates is aromatic extraction unit where the objective is to achieve mainly VI property of LOBS though in this process, CCR, aromatic content and TAN are also controlled partly. Solvent used in this process can be either furfural or N-methyl pyrrolidone (NMP); if furfural is used, the unit is called furfural extraction unit (FEU) and if NMP is used, the unit is called (NMPEU).

The aromatic extracted distillates as mentioned above are then processed in Solvent dewaxing unit (SDU) where the property like pour point of LOBS is achieved.

Though color, color stability, and oxidation stability are partly achieved in FEU or NMPEU, to meet the specification of LOBS, the dewaxed oil (DWO) obtained from SDU is processed in mild hydrogenation unit called a hydro finishing unit (HFU).

As discussed above, vacuum distillates, viz. SO, LO, IO, HO, and DAO (obtained from PDAU) are used in this unit separately as feedstock in different times in blocked out operation mode decided by the planning department of the refinery. The process/technologies are described as follows.

2.4.1.1 Propane De-asphalting Unit (PDAU)

- **Principles of de-asphalting:**

  It is a liquid-liquid extraction process using liquid propane as solvent and vacuum residue as feedstock to the unit. Propane at ambient temperature liquefies at 12 bar pressure. In PDAU, liquid propane is used as solvent for the oil phase of vacuum residue to dissolve the oil and creates extract phase while the remaining of vacuum residue, i.e. asphalt comes out as raffinate phase with lean concentration of propane in it. As the extraction temperature is kept at about 45 to 50°C, extraction pressure can be theoretically kept marginally above 12 bar to keep propane at liquid stage, but to ensure fluid dynamics, i.e. to ensure propane remains in liquid phase all along the process circuit till the outlet of heat exchangers or furnace, the extraction pressure has been found to be kept at minimum above 30 bar. However, now-a-days, ROSE[4] (residual oil super-critical extraction) technology has become popular to reduce energy consumption in PDAU where in the ROSE separator vessel located at downstream of ROSE heat exchanger, the pressure of that heat exchanger and that vessel should be kept above the super-critical pressure of propane mixed with the feedstock, i.e. at temperature of about 105°C, the super-critical pressure of propane comes about 35 bar and above with propane to feedstock dilution ratio of about 8:1 (V/V). Hence in applying ROSE technology in PDAU, the extractor pressure should be kept above 35 bar.

  While the solvent, propane preferentially dissolves paraffin and naphthene compounds of the vacuum residue, it also to some extent dissolves some heavy aromatics present in the residue thereby affecting quality of the de-asphalted oil (DAO), i.e. extract if excessive solvent flow is used; vice versa, if solvent flow is kept low, it would leave some of the oil (paraffins and naphthenes) into the raffinate phase, i.e. in the asphalt thereby reducing the yield of DAO; hence, an optimum solvent ratio is to be used. The extraction temperature also plays similar role like solvent ratio as mentioned above, but propane having reverse solubility in the oil, extract yield would be affected though better in quality when temperature is raised and vice versa, i.e. when temperature is reduced, extract yield would be increased with compromise in quality; hence, extraction temperature also should be optimized in combination with solvent flow ratio. However, contribution of temperature is less than that of solvent ratio in the extraction.

- **The process flow description – a case study:**
  Vacuum residue at a temperature of about 130°C is pumped using positive displacement pump due to its very high viscosity and fed marginally below the top zone of the extractor at a pressure of about 35 bar, and the liquid propane which is on recirculation by its recovery from the system (as discussion follows) is pumped from the condensed liquid propane vessel inside the unit and fed marginally above the bottom of the extractor at a temperature of about 45°C and at a pressure of 35 bar as mentioned above.

  In the extraction process, propane being much lighter moves upwards, and feed being much heavier moves downwards, and in this process, propane dissolves the oil as mentioned above and moves upward to exit from the extractor top called extract phase or DAO-Propane phase but at a very low temperature of about 70°C due to using very high solvent ratio of about 8:1 which helps to decrease the top temperature by cooling the feed at very high temperature of 130°C as mentioned above. The asphaltene compounds of the feedstock being very heavier moves downward along with very less quantity of propane and comes out from the bottom of the extractor as raffinate phase.

  There is installed another small vessel at the top of the extractor where the extract (DAO + Propane) enters through mild heating heat exchanger thereby facilitating precipitation of small quantity of heavier aromatic compounds in that vessel and the refined quality of extract then exits from the top of the vessel. The precipitate in that vessel is pumped back to the extractor as recycle in the extraction process.

  The extract phase leaving from the top of the extractor finally goes to ROSE heat exchanger (after heat exchanging with unit rundown product, DAO in a small heat exchanger) where it is heated up to 105 to 108°C when pressure is about 35 to 38 Kg/cm^2g. At this temperature and pressure, the super-critical separation of DAO from propane takes place, i.e. when the ROSE exchanger outlet enters the ROSE vessel, there would be two phases inside the vessel, i.e. pure propane would come out from the top of the vessel and DAO with very lean concentration of propane would come out from the bottom of the vessel.

  The liquid propane from top of the ROSE vessel enters water coolers and flows to the accumulator vessel for pumping and recirculating into the extractor.

  The bottom of the vessel, lean in propane concentration moves to heat exchangers where medium pressure steam at about 15 bar is used to evaporate the propane after the heated mass moves to a vessel due to keeping the vessel at a lower pressure of about 20 to 23 bar. The evaporated propane at about 70 to 75°C from the top of the evaporator vessel enters to a parallel stream of water condensers where the vapor propane is condensed and moves to the same accumulator as mentioned above.

The bottom of the evaporator vessel, containing DAO with negligible propane, enters a steam stripping column to scrub out the residual propane with help of stripping steam at very low pressure of about 5 bar and temperature of about 150°C thereby facilitating solvent free DAO leaving bottom of the stripper which is then pumped to product rundown tank via a small heat exchanger upstream of ROSE heater as discussed earlier to cool the product to about 80°C.

On the other side, the asphalt-propane mix, i.e. raffinate leaving the extractor bottom as mentioned above, moves to the furnace to heat it up to about 215 to 225°C and then enters a flash vessel where bulk of the propane is evaporated and joins the propane stream ex DAO evaporator and moves to condensers as discussed above. The pressure of the flash vessel remains lower at about 20 bar due to keeping the vessel top connected with the DAO flash vessel top vapors.

The bottom of the asphalt flash vessel moves to a separate steam stripper at about 5 bar and at a temperature of about 180°C similar to discussed above for DAO stripper to strip out the residual propane from asphalt stream, but due to presence of some stearate compounds in the asphalt, there is a foaming tendency in asphalt stripper which makes it difficult to pumping it out from stripper bottom resulting in stripper level rising and thus upsetting operation; to overcome this problem, an antifoam compound dosing is done to asphalt stripper feed inlet which stops the foaming and ensures easy pumping out of the asphalt. The propane free asphalt from the bottom of the asphalt stripper is pumped to asphalt tank via a water cooler to cool the asphalt to about 130°C.

Both the wet propane-water vapor streams coming out respective strippers move to a water cooler to cool up to about 35°C and enters a water knock out vessel to drain the condensed water from it. The propane vapor from the knock-out vessel enters the suction of a reciprocating two stage compressor where propane is pressurized up to about 20 bars by keeping it floating with the overhead system of DAO flash evaporator and asphalt evaporator as discussed earlier above. To keep the propane water free, there is a knock out vessel at downstream of a water cooler after first stage of the compressor wherefrom water if any is drained out.

The schematic flow diagram demonstrating process flows, control, and heat exchangers' integration as discussed above is shown in Figure 2.2.

- **Salient properties of DAO and asphalt:**

To ensure getting quality base stock, i.e. bright neutral category of base oil from subsequent flow processes, viz. FEU / NMPEU, SDU and HFU (or CIDW unit), operating parameters in the extractor and in ROSE circuit are to be maintained as discussed above. The desired properties of DAO are to be achieved along with asphalt for which also quality is important to ensure its blending to produce paving bitumen. The desired properties are given in Table 2.6.

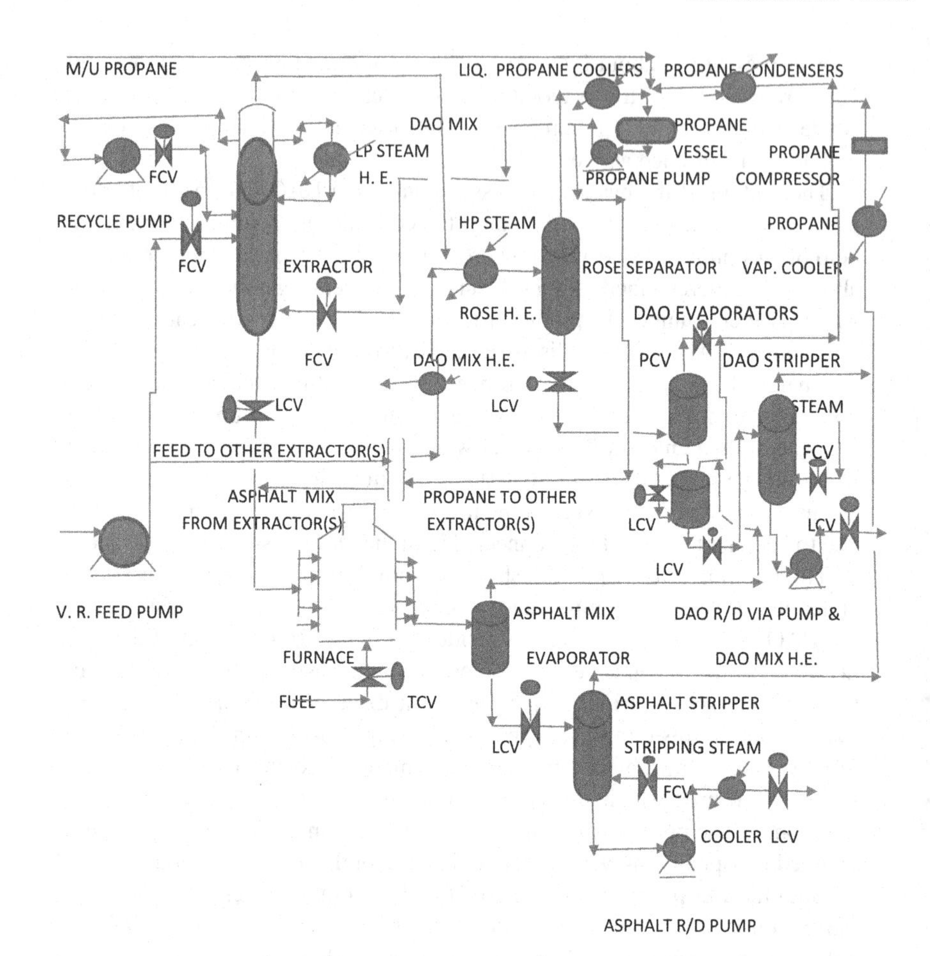

FIGURE 2.2 A case study schematic flow diagram of PDAU.

TABLE 2.6
Desired Properties of DAO and Asphalt and Yield against Feed, VR (Vacuum Residue) Qualities – A Case Study Report

Attribute	Feed (VR)	DAO	Asphalt
Viscosity, cst @98.9°C	1000	33 to 36	------
CCR, %wt.	20–25	2 to 2.5	30–35
Flash Point, deg.c.	>300	>280	>280
Propane content, ppm	---	<10	<10
Penetration	200–300	------	<5
Yield, %wt.	---	25–30	75–70

TABLE 2.7
Case Study Operating Data of Extractor and ROSE Circuit

Attribute	Extractor	ROSE Separator
Solvent to Feed ratio, V/V	7:1 to 10:1	-----
Feed Temperature, deg.c.	110 to 130	-----
Solvent temperature, deg.c.	30 to 40	-----
Top temperature, deg.c.	70 to 78	----
Bottom temperature, deg.c.	45 to 52	-----
Top Recycle to Feed ratio, V/V	0.1:1 to 0.15:1	-----
Pressure, Bar	36 to 40	36 to 39
ROSE temperature, deg.c.	----	105 to 108

TABLE 2.8
Case Study Utilizes Consumption in PDAU

Utilities per MT of Unit Feed Capacity	Expected consumption rate
Propane, Kg	2.5
Furnace Fuel, Kg.	15
LP steam (70 psi), Kg	50
MP steam (200 psi), Kg	100
HP steam, Kg	100
Electricity, KWH	30
Circulating Water, M^3	20

- **Typical operating parameters of PDAU as a case study:**

The operating parameters in the extractor and ROSE circuits are the main operating parameters to be controlled as discussed earlier to get the desired DAO yield and quality while other parameters in the recovery circuits also to be maintained as discussed earlier. But from optimization point of view, extractor and ROSE circuit operating parameters are the areas of concern; hence, the same are given in Table 2.7, providing a range for optimization.

- **Case study utilities consumption:**

The expected utilities consumption in PDAU are shown in Table 2.8.

- **Metallurgy and investment cost:**

Carbon steels are used in this unit. As there are three sections in this unit, viz. high pressure, medium high pressure, and low pressure sections, with temperature ranges normal to moderate high as discussed in process description, and as there are no chemical reactions, carbon steel metallurgy is adequate with varying thicknesses from high to low in these sections. The tubes in the water coolers are generally of admiral T Brass; if capacity of the unit is within 500 thousand tons per annum, then furnace in the asphalt recovery section becomes lower not justifying the economics of providing balanced draft furnace with APH system and instead low-cost induced draft natural convection furnace is used.

With the above, the approximate installed cost becomes 0.2 million USD per TMTA (thousand metric tons per annum) of unit feed capacity for a 500 TMTA plant.

- **Illustrations:**
  - i. Q. CCR of DAO is marginally higher than the desired value, what should be the corrective measure?
    A. Extractor top temperature should be increased by 1 or 2°C which would help in precipitation of asphaltene in the vessel at the top of the extractor which in turn would be recycled down to the extractor by the recycle pump and/or increasing the recycle flow by this pump by 1 or $2m^3hr.$ rate to reduce the CCR of DAO.
  - ii. Q. Either or both viscosity and CCR of DAO are appreciably higher than desired value(s), what should be corrective measure?
    A. Solvent ratio should be decreased by decreasing solvent flow rate marginally by 5% more or less to reduce the solubility of aromatics / asphaltene into DAO.
  - iii. Q. Either or both viscosity and CCR of DAO are in giveaway of qualities, i.e. appreciably less than desired value(s), what should be the corrective measures?
    A. Solvent ratio should be increased preferably by increasing solvent flow rate or by reducing feed flow rate which in turn would increase solubility of heavier oil into DAO thereby increasing both viscosity and CCR of DAO; if CCR was in order beforehand and now exceeded from desired value, then after this flow adjustment, extractor top temperature and/or recycle flow should be marginally increased to the CCR of DAO.
  - iv. Q. If DAO yield is lower than expected value but qualities like viscosity, CCR are normal i.e. at about desired values, what corrective actions can be taken?
    A. Nothing should be done as this situation indicates that there is less DAO potential in the feed (vacuum residue). Following illustration clarifies DAO potential of feed.
  - v. Evaluation of DAO yield/potential:
    Q. DAO with CCR of 2.5%wt. quality is to be produced from VR with CCR of 25%wt. The CCR of by-product, asphalt would be 33%wt. find out DAO yield.
    A. Assume DAO yield at X %wt.; CCR being additive property, the yield of DAO is then given by:

$$(X) \times 2.5 + (100 - X) \times 33 = 100 \times 25$$

i.e. $2.5X + 3300 - 33X = 2500$
i.e. $30.5X = 3300 - 2500$
i.e. $X = 800/30.5$
i.e. $X = 26.23$
Hence, DAO yield would be 26.23%wt.

vi. Q. How the extractor operating pressure and ROSE heat-exchanger temperatures are decided?

A. Equilibrium graph on pressure vs. temperature is available for propane against its purity. For extraction process in PDAU, generally propane with about 98% purity is used as solvent. Now, for this purity of propane, refer to the graph, and find the super-critical pressure and corresponding temperature, called super-critical temperature. Accordingly, extractor should be operated above this pressure but within designed safe values of the extractor, and ROSE heater temperature also accordingly selected as discussed above.

2.4.1.2 Aromatic Extraction of Vacuum Distillates (AEU)

- **Furfural extraction unit (FEU):**

The distillate (SO or LO or IO or HO or DAO) is pumped to the liquid-liquid extractor in this unit from respective storage tank. The feed first enters the de-aerator column which under vacuum removes the air absorbed, if any, in the feed in order to avoid degradation of furfural subsequently; then the de-aerated feed enters the extractor marginally above the extractor bottom, and the solvent, viz. furfural enters the extractor from its top below the top exit of the raffinate. Feed temperature generally remains at about 65 to 80°C depending upon the feedstock used as mentioned above, and solvent temperature varies between 165 to 180°C depending upon feedstock used as mentioned above thereby keeping a temperature gradient of about 100°C between top and bottom of the extractor.

There are simple separation trays like sieve trays inside the extractor. Furfural, having property to dissolve the aromatics into it, flows downward along with absorbed aromatics (called extract-furfural mix or extract mix)) leaving the non-aromatic oils flowing upward to the extractor (called raffinate-furfural mix or raffinate mix) by density difference. Non-aromatic oil has lower density as compared to furfural plus aromatics. Though the aromatics would be furfural rich, non-aromatics also would contain some furfural.

As per extraction phase equilibrium, it is observed that about 90% of furfural remains in the extract and balance 10% goes to raffinate. The yield of raffinate varies from about 70% to 55%wt. from lightest grade (SO) to heavier grade (HO) as aromatic content found increases in the feed from lighter grade to heavier grade. DAO though heaviest among all these grades, its yield is highest as compared to all grades as before this extraction process, major of the aromatics are removed along with asphaltene in PDAU which results in lower aromatic content in DAO thereby contributing higher raffinate yield in FEU.

The extraction process is controlled mainly by temperature as mentioned above and solvent (furfural) to feed ratio while pressure found to have negligible impact in extraction.

Generally, pressure is decided by the hydraulics of the process, i.e. to ensure steady flow of raffinate-mix and extract-mix from top and bottom of the extractor respectively to downstream respective solvent recovery circuit to recover furfural from each stream to facilitate recycle of the same solvent after cooling back to the extractor.

The furfural free raffinate and extract from the respective recovery circuit flows to respective product and by-product storage tank respectively after cooling through heat exchanger exchanging heat with cold stream in the process followed by flowing through water cooling heat exchanger. In the extraction unit flow rate of solvent is generally higher than that of feed, and this solvent to feed ratio increases proportionally from lighter grade to heavier grade feedstock due to aromatic content increases from lighter grade to heavier grade. Though aromatic content in DAO is less than lighter grade as discussed earlier, its solvent ratio is higher due to solvent miscibility point of view with feed viscosity very high as compared to lighter grade along with non-aromatic compounds structural variation as would be seen from data table subsequently.

Due to this higher solvent load, the volume of furfural extract mix in the extractor is much higher than that of furfural raffinate mix, and accordingly raffinate-extract interphase indicator is found near to the top of the extractor. Commensurate with the above discussion, a case study process flow scheme of the extraction process demonstrating stream flows, control and heat exchangers network is shown in Figure 2.3.

From Figure 2.3, it is observed that there are furnaces installed in each furfural recovery section; this is so because the heat and temperature needed for vaporizing of furfural from respective raffinate or extract can't be met by only exchanging heat from hot streams through heat exchangers as shown above. As mentioned earlier, that the amount of furfural flow should be much higher in extract-mix than that in raffinate-mix, heat load in the furnace in extract-mix recovery section is much higher; to take care effective recovery of furfural from this section, evaporation is carried out in multi stage evaporator process through using high pressure and temperature in this circuit to take care of heat and mass load unlike in raffinate-mix circuit where single stage evaporation of furfural after the furnace is found adequate and that too at lower pressure and temperature; temperature of evaporation is also restricted in raffinate-mix circuit to avoid boiling of raffinate. However, after two stage evaporation of furfural in extract-mix circuit and single stage evaporation in raffinate mix circuit, the respective extract and raffinate need to be processed in two stage vacuum stripper one without stripping steam and the other with stripping steam for each of these two circuits before routing the respective by-product and product to storage tank to ensure there is no presence of furfural in these two final by-product and product respectively thereby making the unit commercially viable. The bulk streams of the furfural evaporated out from the evaporators as discussed above flow to distillation column called solvent drying column where the lower pressure of the column facilitates removal of moisture (water) vapor from the top of the column with the help of the heat content present in these bulk furfural vapor streams. The bottom of this column, free

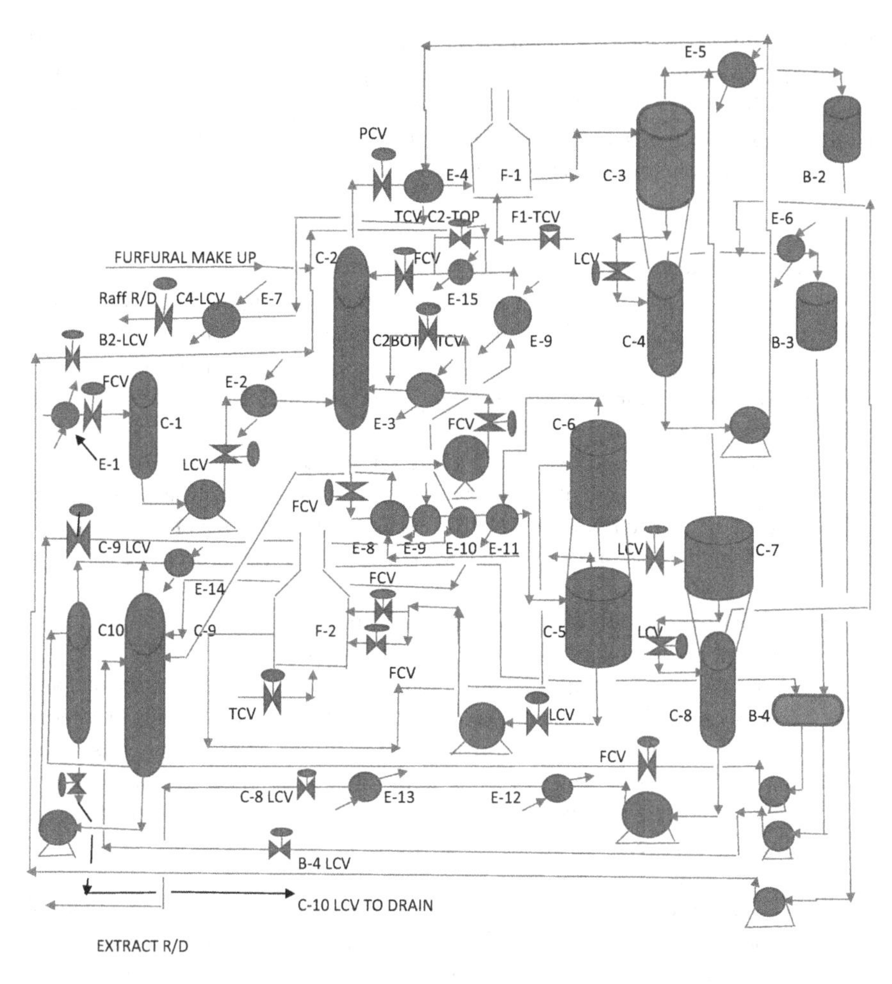

FIGURE 2.3 Case study schematic flow diagram of FEU.

of water, goes back to the extractor after cooling through respective heat exchangers and water cooler as per heat optimized heat exchanger network designed beforehand. The top water vapor from this solvent drying column contains some furfural with furfural / water ratio of about 1:14 due to azeotrope formation equilibrium between furfural and water which means no further separation of furfural from water is possible by distillation though there is adequate difference in boiling point between these two; thus, recovering total furfural again becomes a challenge. To overcome this problem, an azeotropic distillation column is provided where the overhead azeotropic mixture from solvent drying column after cooling through a water cooler (called azeotropic cooler) is fed from the top, and hot stripping steam is introduced from the bottom of this column which ensures some pure water comes from the bottom of this column for open draining though similar azeotrope of furfural-water exits from the top of this column and joins the same azeotrope cooler as mentioned above. But to ensure efficiency of this separation, the furfural-water azeotrope mix first enters a horizontal vessel where through a partition wall where lighter phase overflows to other chamber thereby creating two phases, one phase of water-furfural which is rich in furfural and the other phase which is lean in furfural content; this lean furfural-water flows to the top of azeotropic column and the rich furfural-water phase enters the solvent drying column as top reflux as discussed earlier.

2.4.1.2.1 Metallurgy Used in FEU and Investment Cost

Carbon steel equipment, piping and fittings of different strength / grades are used depending on temperature and pressure in the respective flow circuits of the unit except in wet circuits Monel piping or stainless-steel piping are used, for example, in liquid furfural-water circuits, Monel piping are used, and in furfural-water vapor circuits, stainless-steel piping is used.

As there is no catalyst and reactor is required in this unit and also, as most of the circuits including furnaces are of carbon steel and the pressures in all sections remain in the range of five to ten bar, the relative cost of the unit is lower, for example, approximate installed cost of 0.1 million USD per TMTA of feed processing capacity.

- **Principles of selection of solvent in extraction:**
  - Furfural is selected as solvent due to following reasons:
    Furfural is a polar compound due to unsaturated bonds in this compound and aromatics also have unsaturation in the bond; unsaturates contribute to polarity in the compound molecular structure and from this point of view it is found that furfural is best solvent both in terms of solvent power as well as selectivity to separate aromatics from distillate to produce LOBS intermediates as discussed above.
  - Solvent power:
    Solvent power of furfural is defined as its ability to dissolve how much oil (aromatic plus non-aromatic) in the lube distillate can be dissolved into furfural rich extract phase which of course depends on the temperature and solvent to feed ratio. With increase in temperature, solvent power increases and it is vice versa with decrease in temperature. Similarly, change in solvent ratio also results in the same direction like temperature as mentioned above.
  - Miscibility temperature:
    The extraction temperature is kept generally 5 to 7°C higher than miscibility temperature of the feed with furfural. Miscibility temperature of the feed is defined as the temperature at which feed becomes miscible with furfural at a particular solvent ratio; this means miscibility temperature varies with solvent ratio; at lower solvent ratio, miscibility temperature would be higher and vice versa at higher solvent ratio as explained in Figure 2.4.
  - Selectivity:
    Selectivity is representative of quality of separation between extract and raffinate. In extraction process, there can't be perfect separation between raffinate and extract phase; always there would be some intermixing, i.e. some raffinate (non-aromatic) would be there in extract and some extract (aromatic) would be there in raffinate after extraction process is completed.

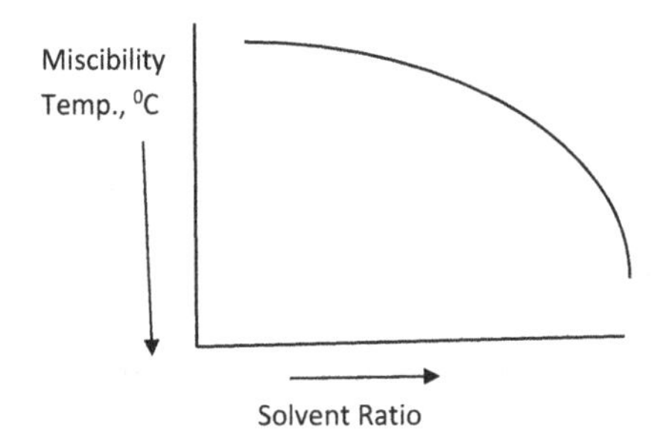

FIGURE 2.4 Miscibility temperature vs. solvent ratio.

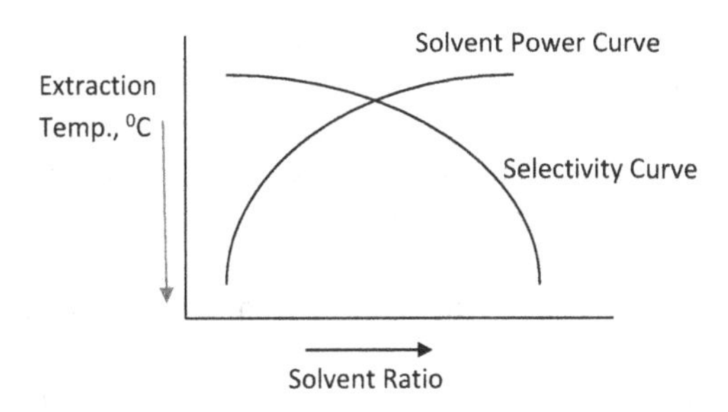

FIGURE 2.5 Extraction temperature vs. solvent ratio.Note: the selectivity follows the above curve at solvent (Furfural) concentration above a minimum threshold value in the extractor.

Poor selectivity is exhibited by two ways – one instance is both extract and raffinate intermix with each other, i.e. practically a very poor separation which happens below a threshold concentration of solvent in the extractor; the other instance is when mainly raffinate goes to extract phase resulting in better quality of raffinate but with decrease in raffinate yield which happens with increase of furfural concentration above that threshold limit or with increase of temperature above that threshold limit. Hence, with change in temperature and/or solvent ratio solvent power changes as discussed above and the selectivity of the solvent acts in opposite direction of solvent power. That's why, solvent ratio and extraction temperature are used to find the optimum solvent power and selectivity, i.e. to achieve optimum yield and quality of raffinate from this extraction process. But below a threshold concentration of solvent, selectivity of furfural gets affected, i.e. major intermixing of raffinate and extract would occur reflected by higher density of raffinate becoming closer to that of feedstock and by lower density of extract becoming closer to that of feedstock as observed in a case study. With ideal separation, the density difference between extract and raffinate should be as higher as practically and commercially possible with feed density in between extract and raffinate.

The above is explained in Figure 2.5.

- Degradability of furfural:

Boiling point of furfural is lower than that of aromatics present in the distillates as discussed above and also adequately higher to follow vaporization at higher temperature and higher pressure.

If proper conditions are maintained, furfural remains stable and does not get degraded.

But care should be taken to avoid furfural slippage from the unit to surface drain going to effluent treatment plant as biological oxygen demand (BOD) of water increases, i.e. dissolved oxygen of water is absorbed by furfural for its degradation thus reducing oxygen content in water thereby threatening the lives of aquatic animals.

Furfural also starts degrading on increasing temperature above 225°C even in absence of air; that's why furnace temperature in furfural-extract recovery circuit is maintained at or below this temperature.

Furfural degrades though at a very slow rate in presence of oxygen as mentioned above even at ambient temperature; that's why in FEU, furfural vessel is kept under inert gas blanketing to avoid ingress of oxygen.

- **Principles of interphase in extractor and controlling extraction:**

As furfural is used as the solvent for aromatic extract, with feedstock being introduced at a zone near to the bottom of the extractor, and as raffinate being lighter than feed exits from top of the extractor, and as extract being heavier than feed exits from bottom of the extractor, and that concentration of solvent (furfural) in the extractor kept adequately high such that raffinate from feed entry zone while traverses upward across the reactor being lighter than feed, there should be enough furfural concentration in the reactor, with furfural entering at the top of the reactor, so that while flowing down due to its higher density than feed, it will purify the raffinate continuously.

With this concept, there would be an interphase indicator between furfural rich extract phase and furfural lean raffinate phase, and this interphase indication is observed near to the top of the extractor. That's why, at the top of the extractor, there are 4 to 5 numbers of sample drawing tapings where by collecting samples from these tapings, one can understand the location of interphase. Some designers provide level indication instrument to know this interphase level, but in many a case, this indicator does not show any level; so physical sample drawing and observation is best approach to confirm if there is zero level indication or not as per interphase level indicator.

Even without taking physical samples as mentioned above when interphase level indication is zero, there is mathematical analysis also by which it can be concluded whether a minimum furfural concentration in the extractor is there or not to avoid intermixing of raffinate and extract exiting the extractor. It is explained as follows.

As explained earlier, with better separation, there would be more density difference between raffinate and extract; hence, accordingly, the pressure differential between top and bottom of the extractor would be higher if there is proper inventory of furfural inside the extractor. It is also explained through the Figure 2.6.

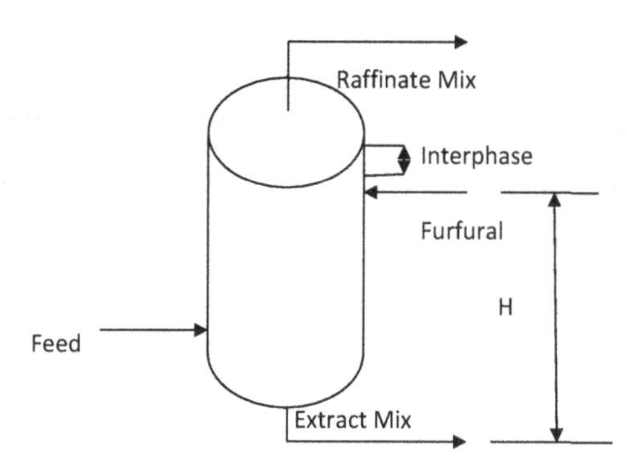

FIGURE 2.6 Phase separation in extractor.

Note: at higher value of 'H', the quality of the raffinate would be better but its yield may be lower.

Ideal extract density is marginally above one, say, 1.0015, and feed density is generally at about 0.88 to 0.95 depending upon lighter to heavier grades. Furfural density is 1.15. So, with adequate furfural inventory, i.e. with furfural concentration more than that of feed in the extractor, the density of the liquid from the bottom up to furfural entry zone would be at least higher than 1.0. Also, pressure at extract mix outlet is represented by:

$$P = H.D.G \tag{2.4}$$

Where H is the height as shown in the Figure 2.6, D is the density of the liquid within zone H above, and G is the gravity constant.

Knowing the value of H, if the pressure indication reading at the bottom of the extractor is found less than the value of P considering value of D => 1, then it can be concluded furfural inventory in the extractor is less than threshold limit and there would be poor extraction as discussed earlier. In such a situation, furfural inventory is to be increased till the desired pressure at extractor bottom is reached.

- **Contribution of temperature gradient in the extractor:**
  - Temperature gradient across the height of the extractor is also another parameter to ensure quality of the raffinate, viz. density which is also reflected by lower value of RI (refractive index) as compared to that of feed while density and RI of extract would be higher than those of feed respectively. For a particular feedstock, desired values of RI of feed, raffinate and extract are known as per experience with a given quality of feed. With higher temperature gradient, better quality of raffinate can be achieved which, however, should to be optimized with respect to yield of the raffinate as discussed earlier.

- **Typical Operating conditions of FEU – a case study report:**

Most important operating parameters are in the extraction process and the same are shown in Table 2.9.

TABLE 2.9
Case Study Extractor Operating Parameters in Furfural Extraction Unit

Sl. No.	Parameter	150 SN	500 SN	150BS
1	Feed: Solvent Ratio, V/V	1:1	1:1.2	1:1.5
2	Feed temp., Deg.C.	80	80	80
3	Solvent temp., Deg.C.	110	115	135
4	Extractor Bottom Temp., Deg.C.	70	75	85
5	Extractor Top Temp., Deg.C.	105	115	130
6	Extractor Top Pressure, Kg/cm^{2}g	2.5	2.5	2.5
7	Extractor Bottom Pressure, Kg/cm^{2}g	5.5	6.5	6.0
8	Extractor Recycle Ratio, V/V	2.5:1	2.5:1	3:1

- **Utility consumption in FEU:**
  Utilities like low pressure (LP) steam (5 bar pressure and 150°C temperature) for piping heat tracing and in heat exchanger(s), medium pressure (MP) steam (15 bar pressure and 280°C temperature) for heat exchanger(s) and in furnace and vacuum ejector circuit, fuel oil for furnaces, cooling water for water cooling heat exchangers, and electric power in the drives of the pumps are used. Utility and chemical consumptions are lower in extraction processes compared to solvent dewaxing and reaction processes, and the same for FEU are given as follows.
- **NMP Extraction:**
- **Principles:**
  The process of aromatic extraction of distillate with NMP as solvent is commonly called NMP extraction and the process unit is called NMP extraction unit (NMPEU). Here, concept is same as that of FEU except in NMPU there is no azeotrope is formed between NMP and water, and thus there is no requirement of using azeotrope distillation column here unlike in FEU as discussed earlier. Regarding solvent power and selectivity, furfural is having marginally higher selectivity than NMP while in solvent power point of view it is vice versa. Also, degradability problem of NMP is less than that of furfural and also from effluent concern point of view, furfural has BOD threat issue while NMP has COD (chemical oxygen demand) issue if there is slippage of the solvent into ETP (effluent treatment plant).

- **Typical operating conditions of NMPEU – a case study report:**

Like in FEU, in NMPEU also, extraction operating parameters are most important as experienced in plant operation and the same are shown in Table 2.10.

- **Utility consumption in NMPU:**

Like FEU, here also, LP steam for piping heat tracing and in heat exchanger(s), MP steam for heat exchanger(s) and in furnace and vacuum ejector circuit, fuel oil for

TABLE 2.10
Extractor Operating Parameters in NMP Extraction Unit

Sl. No.	Parameter	150 SN	500 SN	150BS
1	Feed / Solvent Ratio, V/V	1:1	1:1.2	1:1.5
2	Feed temp., Deg.C.	80	80	80
3	Solvent temp., Deg.C.	110	115	135
4	Extractor Bottom Temp., Deg.C.	70	75	85
5	Extractor Top Temp., Deg.C.	105	115	130
6	Extractor Top Pressure, Kg/cm^2g	2.5	2.5	2.5
7	Extractor Bottom Pressure, Kg/cm^2g	5.5	6.5	6.0
8	Extractor Recycle Ratio, V/V	2.5:1	2.5:1	3:1

TABLE 2.11
A Case Study Utility Consumption in FEU and NMPU

Utilities per MT of Unit Feed Capacity	FEU	NMPEU
Solvent, Kg	1.25	1.0
LP steam (70 psi), Kg	50	50
MP steam (200psi), Kg	100	80
Fuel, Kg	20	18
Electricity, KWH	15	15
Circulating Water, M^3	14	14

Note: the figures are for 500SN grade feed; for BS grade feed, utilities consumptions are 30% higher due to feed rate lower by 40% which can save only 10% utilities due to same unit is used for operation at lower feed rate while designed to operate at higher feed rate.

furnaces, cooling water for water cooling heat exchangers, and electric power mainly in the drives of the pumps are required.

The utility and solvent consumptions in FEU and NMPEU are given in Table 2.11.

- **Desired properties for raffinate and by-product, extract achieved in FEU and NMPEU:**

Whatever may be the solvent used, either furfural or NMP, the quality of raffinate and extract remain same except variation in the processes and utilities consumption as explained. The desired properties to be achieved in FEU and NMPEU in order to meet the ultimate specifications of 150N, 500N and 150BS from HFU are shown in Table 2.12.

- **Metallurgy and investment cost:**
  Carbon steels are used in this unit. As there are four sections in this unit, viz. feed extractor, extract-solvent mix recovery, raffinate-solvent mix recovery and

TABLE 2.12
Expected Properties of Raffinate and Extract and Yield Based on Feed Qualities

Sl. No.	Attribute	150N	500N	150BS
1.	Raffinate:	4–5	8–9	29–31
	-Viscosity, cst. @100°C	0.85–0.87	0.87–0.89	0.89–0.91
	-Sp. Gravity	0.01–0.02	0.03-.04	0.8–0.9
	-CCR, %wt.	1.464–1.466	1.467–1.469	1.480–1.482
	-Refractive Index (RI)	68	60	65
	-Yield, %wt.			
2.	Extract:	0.87–0.88	1.00–1.02	0.98–0.99
	-Sp. Gravity	1.525–1.530	1.561–1.565	1.540–1.545
	-Refractive Index (RI)			
3.	Feed (Distillate for 150N and 500N and	5.5–7.5	9.5–11.5	31.5–33.5
	DAO for 150BS):	0.87–0.88	0.89–0.90	0.91–0.92
	-Viscosity, cst @100°C	1.475–1.480	1.500–1.510	1.485–1.490
	-Sp. Gravity			
	-RI			

Note: extract density and RI of 150 BS lube equivalent raffinate are lower and yield of raffinate is higher than those in 500N case, because much of the aromatic and asphalt are removed in PDAU while producing DAO.

solvent drying sections, all being low pressure circuits with pressure varying from two bar to three bar abs. maximum including vacuum sub-sections, with temperature ranges normal to moderate high as discussed in process description, and as there are no chemical reactions, carbon steel metallurgy is adequate with conventional lower thickness except a bit higher in the furnaces. The tubes in the water coolers are generally of carbon steel except using SS tubes for only a few exchangers and Monel piping in a small circuit where there is a possibility of corrosion by furfural-water system; if capacity of the unit is within 500 thousand tons per annum, then furnace in the asphalt recovery section becomes lower not justifying the economics of providing balanced draft furnace with APH system and instead low-cost induced draft natural convection furnace is used.

- With the above, the approximate installed cost becomes 0.15 million USD per TMTA (thousand metric tons per annum) of unit feed capacity for a 500 TMTA plant.
- **Illustrations:**
  i. Q. Raffinate CCR and/or RI is higher or lower than normal; what should be corrective measure?
     A. If it is marginally higher, then increase the top temperature of the extractor by 1 or 2°C, and/or reduce the bottom temperature of the

extractor, i.e. to maintain the desired temperature gradient between top and bottom of the extractor; also, bottom recycle flow increase can be tried to see any improvement in quality of raffinate and/or extract; but if it is appreciably higher, then increase the solvent ratio preferably by increasing the flow rate of the solvent or by reducing the feed rate; then, if necessary, adjust the top temperature for fine tuning of the qualities as mentioned above.

If CCR and/or RI are lower than normal, it means there is quality give-away; in that case, reverse action(s) should be taken to reduce / minimize quality give away.

ii. Q. Raffinate CCR and RI are appreciably higher than normal, the above actions are taken, but there is no quality improvement, what should be next course of action(s)?

A. The above situation indicates that there is not minimum inventory of the solvent inside the extractor, and in that case, there would be no visible interphase level indication at the top of the extractor. Note, extractor is so designed that interphase level indication should be visible near the top zone of the extractor where the liquid above that level is called raffinate-solvent mix and liquid below that level is called extract-solvent mix, and while feed enters near bottom of the extractor, the raffinate-solvent mix moves upward and gets purified throughout the whole length of the extract-solvent mix phase till it reaches up to the interphase.

The non-visibility of the interphase level can also be counter-checked by observing the pressure at top and bottom of the extractor; in the industry, the top pressure remains at about 2–2.5 Kg/cm^2g and bottom pressure remains at about 5.5–6.5 Kg/cm^2g, i.e. there is normal differential pressure of about 3.5 to 4.0 Kg/cm^2g depending upon extractor height when there is enough solvent inventory inside the extractor. When there is not adequate solvent inventory inside the extractor, this differential pressure would be lower accordingly.

In such situation, first action should be to receive solvent into the extractor to get back the desired solvent inventory as mentioned, then the action(s) as mentioned in illustration (i) should be followed.

iii. Q. How to evaluate raffinate yield:

Intermediate viscosity distillate is processed in FEU to produce raffinate with ultimate objective to produce 150N LOBS. The RI of feed, raffinate and extract are 1.500, 1.468 and 1.560 respectively. Find out yield of the raffinate.

A. RI is an additive property for the feed source (distillate/DAO) obtained from same crude source. Also, the additive property would be applicable in a steady state of extraction, i.e. when an interphase level is visible in the extractor. Assuming the extractor operation is in steady state as mentioned above, the raffinate yield assuming it to be X would be given by:

$1.500 \times 100 = (X) \times 1.468 + (100 - X) \times 1.560$
i.e. $150 = 1.468 + 156 - 1.56X$
i.e. $(1.56 - 1.468) X = 156 - 150$
i.e. $0.092X = 6$
i.e. $X = 6/0.092$
i.e. $X = 65.2$,
i.e. raffinate yield is 65.2%wt.

iv. Q. Which is the better solvent among furfural and NMP for aromatic extraction of lube/wax bearing vacuum distillate derived from petroleum crude oils?

A. The advantage of using furfural is that it is more selective than NMP while solvent powers are comparable, i.e. better quality of raffinate can be achieved by using furfural instead of NMP, but disadvantage is that furfural is degradable during storage in tank due to ingress of oxygen, if any, unlike NMP; its slippage into effluent water, if any, increases BOD (biological oxygen demand) of the effluent water thereby causing threats to aquatic animals in the water unlike NMP. Another disadvantage of using furfural is that it gradually decomposes /degrades into decomposition products at higher temperature above about 225°C through auto-catalytic reaction in presence of iron or rust particle unlike NMP, and the fact is that the furnace outlet temperature in solvent-recovery circuit remains at about 220–225°C which in turn makes the furnace tube metal temperature (called the skin temperature) to be as high as 350–400°C thereby facilitating slow decomposition of furfural unlike NMP resulting in net furfural make up consumption unlike NMP.

Hence, it's a trade-off by the operating people to decide the use of the solvent.

2.4.1.3 Solvent Dewaxing of Raffinate Obtained from Aromatic Extraction

Raffinate so obtained from FEU contribute to VI improvement of the lube distillate as discussed earlier, but the waxy paraffin present in the raffinate causes congealing problem in LOBS unless these waxy paraffins are not suitably removed. The objective of the solvent dewaxing unit (SDU) is to separate out this wax from the raffinate to remove this congealing problem as reflected by lower pour point after processing through this unit.

We know aromatics have lower VI, hence, after aromatics removal in FEU, the VI of raffinate increases from the level of about 55 in the feed to FEU to the level of about 105 in the raffinate. Similarly, among the non-aromatics, paraffins have higher VI. In SDU, all paraffins are not removed, only waxy paraffins are removed to improve pour point. Hence, after processing raffinate through SDU, VI of the dewaxed oil (DWO) reduces partially from the level of 105 to the level of 95 which is adequate to use the product as LOBS.

The process/technology of solvent dewaxing is guided by following principles:

- **Selection of solvent and refrigeration/crystallization:**

Main concept is refrigeration of the raffinate to ensure precipitation/crystallization of waxy paraffins and then filtering the chilled mass to separate out the wax. If the process is carried out without use of any solvent, the quality, viz. pour point of DWO will be the best but the yield of DWO would be very less with lots of desired oil passing into wax phase, also there would be fluidity constraint in the process to carry it out hydraulically; thus, the process would not be commercially viable.

Role of solvent:
Solvents like benzene and toluene had been found to be most effective solvents to the non-waxy part of the raffinate in the descending order of aromatics, naphthene, naphthene with paraffins in the side chains, and waxy paraffins. But benzene, now-a-days, is no longer used as solvent due to its carcinogenic property. But both toluene and benzene have some solubility to wax also after dissolving the oil into it. On the other side, some solvents have been found to be very effective as anti-solvent to waxy paraffins in addition to its high solvency power to the oil, that means it throws away the waxy paraffins while it comes into contact with non-waxy components as mentioned above. Best example of anti-solvent used in SDU is MEK (methyl ethyl ketone). But if excessive MEK is used in the solvent, then after throwing away waxy paraffins, it starts throwing away non waxy components also thereby creating an oil layer in the separated wax phase after crystallization. It has been found that an equal mix of toluene and MEK (methyl ethyl ketone) is the best solvent to be used for this purpose.

Rate of crystallization is one of the most important parameters in SDU for effective performance of the process, i.e. it decides the perfect phase separation between waxy and non-waxy oils of the raffinate being processed. Rate of crystallization is defined by the following expression:

$$C_R = K_1 . (C - S) / \mu \quad (2.5)$$

Where C_R is the rate of crystallization; K_1 is a constant; C is concentration of the solute, viz. waxy paraffins, S is the solubility of wax in the solvent and μ is the viscosity of the total mass (raffinate plus solvent) where solvent means solvent plus anti-solvent, i.e. MEK plus toluene.

Here both C and μ depend upon the solvent dilution ratio, i.e. with higher solvent dilution, value of C and μ would decrease and vice versa. Hence, from the above equation, it is seen that with higher solvent dilution ratio, rate of crystallization would decrease and vice versa. So, in SDU, an optimum solvent dilution ratio is to be established for each grade of raffinate processing.

The above equation is a simple one and is in use in the industry; there are also other different correlations connecting these variables to predict the crystallization rate more accurately.

- **Solvent ratio and crystallization temperature:**

Crystallization temperature is defined as the temperature at which the waxy paraffins start crystallizing. Here the process, crystallization, includes first formation of a wax crystal (crystal nuclei) on which gradually other wax crystals are formed and accumulated on further chilling thereby controlling the crystal growth; the crystal growth terminates after certain residence time even without further chilling. Still chilling further would result in crystallization of the non-waxy paraffins affecting DWO yield.

Crystallization temperature of waxy paraffins in the raffinate can be predicted by the cloud point and pour point of feedstock, raffinate through correlation.

The crystallization temperature, $T_{C,}$ can simply be expressed by the following equation:

$$T_C = K_2 . C_S \tag{2.6}$$

Here, the value of C_S will decrease with increase in solvent dilution ratio resulting in decrease in crystallization temperature and vice versa.

From the above two equations, it is evident that an optimum solvent dilution ratio is to be established to achieve a techno-commercially viable crystallization temperature and rate of crystallization.

Other parameters to control the crystallization:

It has been found by the technology houses that to control the crystallization with respect to achieve higher DWO yield along with its quality (lower pour point), solvent dilution should be done at various stages of flow of the feed during the progress of chilling and the dilution ratio should be optimized in each stage. First solvent dilution done at the feed entry at ambient temperature or above in the unit is called primary dilution, after certain stage of chilling, say, at about 20°C or at about 15°C, further solvent dilution done is called secondary dilution; sometimes after crystallization is complete, some more solvent dilution is done before filtration which is called tertiary dilution.

In addition to the above, to control crystallization, chilling rate also should be optimized; it has been established that chilling should be done gradually instead of doing rapidly. In rapid crystallization, some shock crystals are formed which are not good nuclei to propagate the crystallization without entrapping non-waxy oil inside the crystals. It is found that to reach chilling temperature up to 20°C from 45°C, chilling rate should be about 2°C per minute and subsequently, the chilling rate can be increased to 4°C per minute.

- **Filtration of the wax-oil slurry after crystallization:**

Unlike in aromatic extraction unit where there is clear interphase formed between raffinate and extract, in SDU, the wax crystals formed is found to be homogeneously mixed with the non-waxy oil and forms a slurry preventing the two phases to be separated without filtration. The filtration is done with the help of rotary vacuum filter where crystals form a cake on the filter cloth and the DWO plus solvent pass through the filter cloth and enter numbers of conduit connected to the filter cloth underneath.

These conduits lead to a vessel to collect DWO plus solvent (called filtrate) with the help of vacuum as the vessel top is connected to a vacuum compressor; to ensure flow dynamics of the compressor, inert gas flows to the filter over the filter cloth which in turn is sucked to the filtrate vessel along with the chilled liquid filtrate; the inert gas exits from the top of the vessel through a pipe to enter the compressor suction. The rotary part of the filter is called the filter drum and the filter drum is housed completely within a stationary closed cylindrical vessel split into two halves with bottom half called the filter vat and top half called the filter hood. The wax cake (wax plus solvent mixture) deposited on the filter cloth moves down during its rotation to a connected vessel (called boot) with the help of a scrapping blade (called doctor's knife) kept stationary on the surface of filter cloth.

While following the above filtration circuit as explained, to achieve good filtration rate and filter cake quality to achieve in turn good yield of DWO and capacity of operation of the unit, one should understand the principles of filtration as follows.

Filtration rate is a function of differential pressure between filter hood and conduits below filter cloth, filter drum speed, filter submergence, i.e. percentage of the filter rotary drum submerged into the filter vat where pool of chilled feed slurry continuously enters from the terminal crystallizer as discussed earlier; this chilled mass forms a level in the filter vat where depending on the level of the vat, a percentage of the drum gets submerged and thus the drum during its rotation picks up the slurry to carry out the filtration as discussed above.

The filtration rate, F_R, is given by:

$$F_R = K\,(\text{Del.P}^{0.84}.\text{R.S.T}^{0.13})^{0.5} \tag{2.7}$$

Where, K is filtration constant, Del.P is differential pressure, R is drum speed in r.p.m. (revolution per minute), S is submergence as a fraction of total vat volume, T is thickness of the filter cake.

By optimizing the above parameters, one can optimize in turn the filtration rate. Here, Del.P is guided by T and also by the nature of filter cake, viz. even for fixed cake thickness, if wax crystals are very small, say, micro-crystalline, then porosity of the cake would be very less thereby causing higher pressure drop for the filtrate to flow across the filter. Also, if cake thickness is high, pressure drop also would be high as usual. Hence, in the process, the objective should be to optimize the solvent dilution to optimize filter cake thickness which in turn will optimize the filtration rate thereby not only increase unit capacity but also increases DWO yield. However, excessive solvent dilution can cause deterioration of DWO pour point due to melting of some wax into the filtrate. Also, after following above to control wax cake thickness, further effort can be made to reduce wax cake thickness by spraying chilled solvent through various nozzle spray pipes placed at the filter hood over filter cloth. There is another factor in Del.P which is porosity of the filter cloth; like these, Del.P is sum of various parameters as reproduced below:

$$\begin{aligned}\text{Total Del.P} = \text{Del.P due to (} &\text{Filter cloth porosity + Filter Cake porosity} \\ &\text{+ Filter cake thickness + Filtrate velocity through cake} \\ &\text{and conduits + Filtrate Viscosity)}\end{aligned} \tag{2.8}$$

In the filtration in SDU, it is found that most critical parameter, among the above in the Equation 2.8, is filter cake thickness and its porosity which cause the major pressure drop. The pressure drop across this filter cake can be calculated using Ergun equation as followed in fluid dynamics engineering by considering the pores of the wax crystal as conduits for flow of filtrate through these; the porosity can be considered as average pore diameter and this filter cake can be considered as packed bed column. Now, Ergun[4] equation can be applied as follows:

$$f_p = 150/(Gr_p) + 1.75 \quad (2.9)$$

where f_p and Gr_p are defined as:

$$f_p = \text{Del. P (thro' conduits)} \times D_p \times (e^3/(1 - e))/(L \times d \times v_s^2) \quad (2.10)$$

and

$$Gr_p = d \times v_s \times D_p /((1 - e)\ \mu) = R_e / (1 - e) \quad (2.11)$$

where:

Gr_p is the modified Reynolds number, f_p is the friction factor of the packed bed of conduits, Del. P (through conduits) is pressure drop across the packed bed, L is the length of the packed bed (filter cake thickness), d is density of the filtrate, μ is dynamic viscosity of the filtrate, v_s is the superficial velocity of the filtrate (considering there is no restriction in the conduits), e is the void fraction (porosity of the bed, i.e. filter cake porosity), and R_e is particle Reynolds number.

- **Process description of SDU – a case study:**

The raffinate from storage tank is lined up with unit feed pump unit at a temperature of about 80°C so that the feed becomes homogeneous and there is no layering of waxy paraffins in the feed. The solvent pump in the unit receives primary dilution solvent from the respective vessel in the unit at ambient temperature and then pumps it to feed pump discharge. The feed and solvent temperature reaches in between 80°C and ambient; in this process, there may be shock chilling of the feed to develop some **shock** crystal, that's why, this feed-solvent mixture is subsequently heated through a heat exchanger to a temperature of about 70 to 75°C, i.e. approximately 10°C above miscibility temperature (similarly established in SDU like Figure 2.4 in FEU), then it is subsequently cooled through a water-cooling heat exchanger to a temperature of about 50 to 55°C to start the gradual cooling for crystallization process. It is found that nucleation of wax crystals starts at the cooler. Next the water-cooled feed-solvent mixture enters tube sides of a series of scrapper chillers (double piped heat exchangers) for gradual chilling of the feed-solvent mixtures. In the first set of scrapper chillers, it gets chilled up to about 20°C by the cold filtrate (obtained from the filters as mentioned earlier) followed by chilling in the next series of scrapper chillers where refrigerant is there in the shell side to chill the feed-solvent mixture up to about 10°C followed by further chilling in next two series of similar set of scrapper chillers to chill the feed-solvent mixture to about

(–)5°C and (–)15°C respectively. In order to maintain fluidity of the feed-solvent mixture across the chillers, secondary dilution solvent is added either at the entrance of first or second set of refrigerant chillers as mentioned above. Also, to avoid congealing of the feed-solvent mixture in the scrapper chillers, there is a rotating blade at a speed of about 10–15 RPM (revolution per minute) inside each tube of the chillers, which scrap out the solid deposit, if any, on the tube inside surface thereby facilitating fluidity of the feed-solvent mass. A chiller set generally contains (4+4) or (6+6) or (8+8) numbers of double piped heat exchangers in parallel decided by the volumetric flow rate of the feed-solvent mixture to get the desired chilling rate as discussed earlier. Chiller tubes are laid horizontally but with about 2°slope downward to facilitate proper draining of the content to slop lines before idling the unit.

Filtration Section:

The chilled mass from the last chiller set outlet enters the rotary vacuum filter as discussed earlier via a vertical surge drum located at elevated position above the filters. In some cases, additional cold dilution solvent is added into the chilled mass again to achieve better filtration. This dilution is called tertiary dilution. The chilled mass enters into number of filters in parallel depending upon feed-solvent mixture flow rate as generally standard capacity of filters are available in the market to be installed economically. Filtration process has been described earlier; filtrate moves to a common filtrate vessel and wax slurry gets collected in the respective filter boots. As the solvent plus anti-solvent have the solvent power to dissolve oil (non-paraffins) as discussed earlier as against paraffins, particularly waxy paraffins, the majority of the solvent plus anti-solvent would be in the oil phase. It is found that about 85% of solvent plus anti-solvent remains in the oil phase and balance about 15% remains in the wax phase decided by the oil-solvent and wax-solvent solubility equilibriums which have been found to be around 90:10 (solvent: oil) and 50:50 (solvent: wax) both vol./vol. respectively. These ratios and with experienced DWO and wax yield of 75% and 25%wt. respectively in turn result in solvent plus anti-solvent distribution of 85% and 15%wt. in oil and wax phase respectively as mentioned above. The filtrate from the filtrate vessel is pumped subsequently to a solvent recovery circuit where through a series of heat exchangers network, and using distillation in phases, solvent is recovered through columns' overhead condensers and collected in solvent vessel and solvent free DWO is pumped from last column bottom, heat exchanged to cool down before sending it to storage tank.

Similarly, wax slurry from each filter boot is pumped by respective pumps, heat exchanged in a series of heat exchangers network, and using distillation in phases, solvent is recovered through respective columns' overhead condensers and collected in the same solvent vessel as mentioned above. The process descriptions of these two solvent recovery circuits are discussed in detail as follows.

2.4.1.3.1 DWO-Solvent Recovery Circuit

The filtrate enters the first distillation column at a temperature of about 110°C with column top pressure marginally above atmospheric pressure (without any pressure control valve at the overhead) just to ensure flow of overhead condensed solvent flows down to solvent vessel located at the ground level. The overhead solvent would be rich in MEK due to its lower boiling point of 78°C as compared to toluene having boiling point of 110°C. The bottom of the column thus would be a mixture of DWO and toluene rich solvent. This column bottom is then pumped again through series of heat exchangers to the second distillation column, but there may not be adequate heat source in the system to heat the first column bottom liquid up to about 170°C before sending it to second column to be operated at a higher pressure of about 2 Kg/cm^2g (with pressure control valve placed at overhead condensers downstream) in order to ensure evaporation out of most of the toluene in DWO-solvent mixture. Hence, either some heat exchangers with steam as heat source or fuel heater is installed to supply the desired heat to achieve such temperature at elevated pressure. The temperature and pressure as said are decided so based on solvent concentration and composition in DWO-solvent mixture and toluene boiling point. The overhead vapor from this column gets condensed in a series of condensers, and the condensed solvent rich in toluene flows down to a second solvent vessel at the ground. Next, the bottom of this second column with lean concentration of solvent is pumped to a third column keeping its pressure reduced to just above atmospheric like that in first column without any pressure control at the overhead of this third column to ensure evaporation of rest of the solvent, but the DWO-solvent mixture is to be preheated through a steam heater beforehand to supply the heat of evaporation. The overhead solvent vapor though may be rich in toluene contains very less solvent than that coming out from the second column. Hence, as the pressure of the third column is nearly same as that of first column, the vapor from its overhead joins the same condensers of first column overhead. Next, the bottom of the third column would still have some solvent though negligible but not acceptable from point of view of DWO quality as well as wastage of solvent; also, at the same time, no more use of evaporating column would be successful. Due to this reason, the bottom of the third column is pumped to a fourth column having stripping steam facility as we know that at lowest concentration of any component, best technique is steam stripping to scrub out that component to take it to overhead of the column. As no further heating and pressure raising are required before sending the third column bottom to fourth column, hence, the third column is placed over the fourth column such that bottom of the third column flows down to fourth column by gravity. The overhead solvent coming out from the fourth column would contain moisture/water due to use of steam stripping in there, and it is also found that MEK of the solvent forms azeotrope with water at MEK:water concentration of about 11:1 and below. Hence, the overhead vapor from fourth column enters a separate condenser (called azeotropic condenser) and the condensate flows down to a separate vessel at the ground level; in this vessel, solvent rich phase overflows from a partition plate to a second chamber of that vessel. So, the vessel should have two types solvent, i.e. in one chamber, the solvent would be lean in water and in the other chamber, it would be rich in water. This rich in water phase is then pumped to another distillation column called an azeotropic distillation column where also stripping steam is used to evaporate out the solvent from it. The solvent-water mixture from overhead moves to same azeotropic condenser as mentioned above and the bottom of this azeotropic distillation column would be free water without any smell of MEK and so, it flows down to open sewage.

2.4.1.3.2 Wax-Solvent Recovery Circuit

Wax-solvent mixture is pumped from each filter boot, flows through a series of heat exchangers before entering first distillation column. This distillation column temperature and pressure are kept at nearly same as that maintained at first column of DWO-solvent recovery circuit due to same reason as discussed earlier. Accordingly, the overhead solvent from this column moves to same condensers connected with the overhead of first column in DWO-solvent recovery circuit. As overall solvent concentration in wax-solvent phase coming out from filters is much less as compared to that in DWO-solvent phase as discussed earlier, there is no need of pressurized heating of the bottom of first column of the wax-solvent recovery circuit before sending it to second column. Hence, only heating is done up to about 160°C before sending it to third column which is operated at near to atmospheric column like that in first and third column of DWO-solvent recovery circuit. Accordingly, the overhead vapor from this second column of wax-solvent recovery circuit enters the same series of condensers as connected to overhead of first and third column of DWO-solvent recovery circuit and first column of wax-solvent recovery circuit before flowing down to a common solvent vessel at the ground level as discussed earlier. The solvent rich water from the azeotropic separator vessel as mentioned above is also pumped to this solvent vessel. The solvent in this vessel accordingly is called wet solvent whereas the solvent coming out to another vessel from second column of DWO-solvent recovery circuit as mentioned earlier is called dry solvent with respect to water. Next to discuss is that, similar to that in DWO-solvent recovery circuit, the bottom of the second column of wax-solvent recovery circuit flows down to third column by gravity for steam stripping of residual solvent from wax-solvent mixture of second column bottom. Also, the overhead of this third column of wax-solvent recovery moves to same azeotropic condenser as mentioned above. As mentioned in the crystallization process earlier, solvent added first at the unit feed pump discharge as primary solvent followed by further solvent added at the exit of first or second refrigerant chiller set (as discussed earlier) as secondary solvent followed by the third phase addition of solvent addition at the exit of final chiller set (which may not be used as per specific design given to particular plant) as tertiary solvent, it may be mentioned that the solvent from wet solvent vessel as discussed above should be used as primary solvent due to its injection at a higher temperature whereas the solvent from dry solvent vessel should be used as secondary and tertiary solvent (where necessary) due to its injection at lower temperature, as discussed above, is chilled in solvent chillers to achieve the desired temperature before injection. To avoid choking in the feed-solvent flow circuit by ice formation due to ingress of water, if any, the free water, if any, from the wet solvent vessel is drained to sewage.

2.4.1.3.3 Chilled Solvent Generation

As discussed in the filtration section principles, chilled solvent is required to be sprayed over filter cloth to reduce wax cake thickness. This chilled solvent is called cold wash. Hence, a part of the dry solvent as mentioned above is sent to a series of solvent chillers using cold filtrate to heat exchange initially followed by using refrigerant chillers (all shell and tube heat exchangers) to generate chilled solvent.

Some chilled solvent from the exit of second refrigerant solvent chiller is taken out as secondary dilution solvent as mentioned above, and the chilled solvent from the exit of final refrigerant chiller is taken as cold wash as mentioned above and a part of it is used as tertiary dilution solvent, if necessary, by the design.

2.4.1.3.4 Solvent Drying Column

We have seen above that primary solvent is wet solvent and there is chance of water ingress in the system from primary solvent; if water draining is not done continuously and completely drained from wet solvent vessel, this water in turn forms ice, in the feed-solvent mixture thereby choking flow in the crystallizers, in cold wash spray nozzles in the filter thereby stopping cold wash flow, and in the filter cake clogging the filter cloth pores to prevent filtration as well as scratching the filter cloth thereby short circuiting of wax through filter and in turn deteriorating the pour point property of end product, DWO. To prevent these undesired phenomena, it is desirable to install a solvent drying column in the unit which is being followed now-a-days. With solvent drying column in position, the flow of solvents from different columns overhead as discussed above would be somewhat different as follows.

The hot solvent vapors from second column of DWO-solvent recovery circuit having highest temperature would enter the drying column near its bottom while the vapors from third column overhead of DWO-solvent recovery circuit and second column overhead of wax-solvent recovery circuit would enter the drying column at an elevated zone and the vapors from first column overhead of DWO-solvent recovery column and vapor from third column overhead of wax-solvent recovery column would enter the drying column at further elevated zone in the order of decreasing temperature of the vapors. Except for these flow path changes, the flow processes in DWO-solvent recovery circuit and wax-solvent recovery circuit are unchanged as discussed earlier. The solvent rich water from the azeotrope vessel as discussed earlier would be now used as a top reflux to the solvent drying column instead of routing it to wet solvent vessel as discussed earlier. In this process, this wet solvent would be dried in the solvent drying column with help of the heat supplied by the hot vapors from different columns entering the drying column as mentioned above. The quantity of hot vapor from top of the solvent drying column would be kept minimum as per heat input provided to the drying column in order to evaporate out mainly water from the column; this vapor enters the azeotropic condenser as discussed earlier followed by moving to azeotropic vessel as discussed. The hot solvent from bottom of the drying column is pumped to the coolers and the coolers outlet solvent at ambient temperature enters both primary and secondary solvent vessel as described earlier but now solvent in both of these vessels would be dry solvent and there would be no more wet solvent in the crystallization and filtration sections of the unit. Including the process of solvent drying, the above process description is demonstrated as a case study by depicting flow streams, process control and heat exchangers network as in Figure 2.7.

2.4.1.3.5 Refrigeration Circuit

Refrigerant used in the chiller set as discussed earlier is generally liquid ammonia (NH_3) for system where chilling of the feed-solvent mixture is to be achieved as

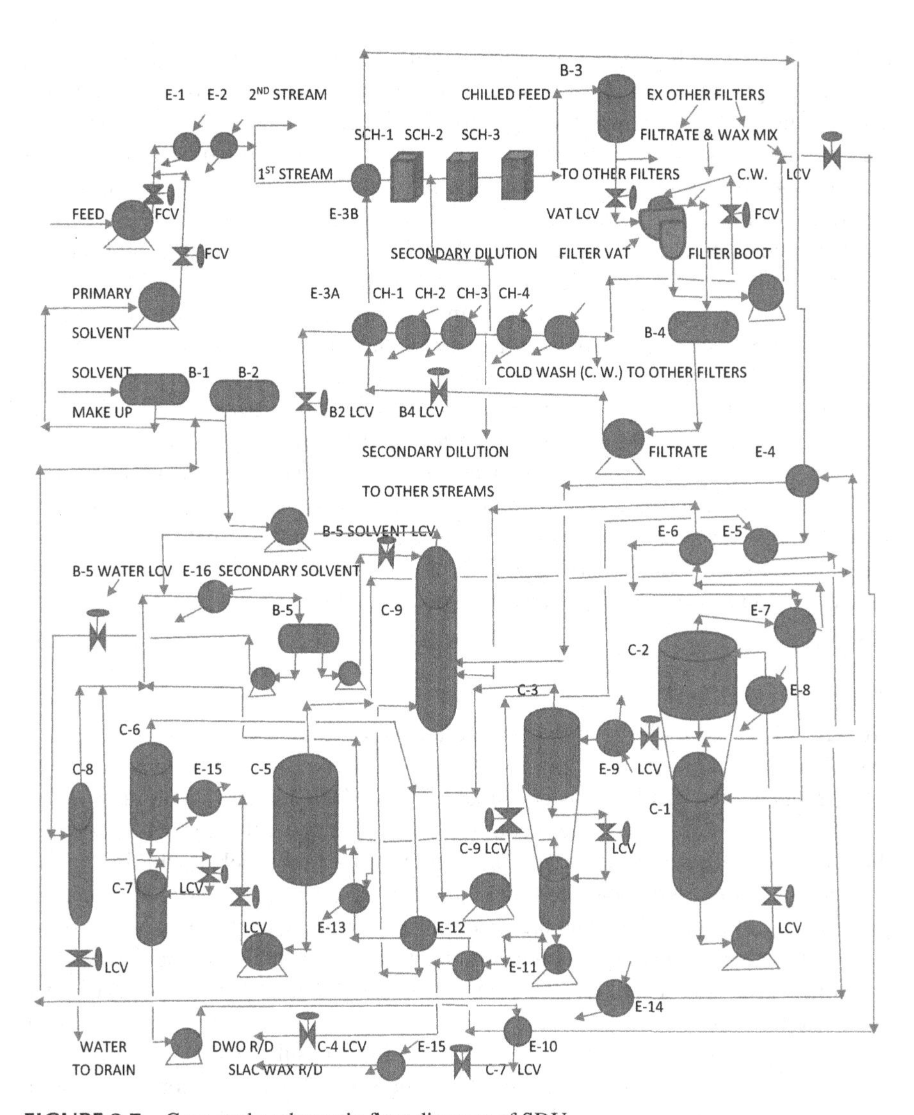

FIGURE 2.7 Case study schematic flow diagram of SDU.

low as (−)15 to (−)18°C. But if ammonia leakage occurs, it causes human inhalation problem; due to this toxicity of ammonia, now-a-days, liquid propylene is also used as refrigerant in solvent dewaxing. In earlier days, ammonia was used as refrigerant. With ammonia as refrigerant, ammonia vapor coming out of scrapper chillers moves to compressor suction vessel. Due to appreciable quantity of ammonia flow, multi-stage centrifugal compressor is generally selected. The operation of such compressor is critical with respect to its successful operation of its lube oil and seal oil system, its discharge temperature, its suction pressure, and its suction vessel level. With lube oil and seal oil high temperature security, compressor trips; with low lube oil and seal oil pressure also, compressor trips. Similarly, with high discharge temperature, high discharge pressure, high suction pressure, and with high ammonia level in suction vessel, compressor trips. Compressor tripping means refrigeration gets affected and the unit is to shutdown consequently. As far as ammonia flow is concerned, the ammonia vapor is compressed from 0.2 bar gauge to about 16 bar gauge in the compressor; the discharge gas at temperature of about 160°C is then condensed to liquid in a water cooler followed by further cooling from 40°C to about (−)10°C in another exchanger exchanging heat with chilled ammonia vapor coming out from the scrapper chillers followed by further cooling to about (−)25 to (−)30°C in another shell and tube exchanger where main ammonia stream at (−)10°C enters tube side and a slip stream of that ammonia is taken to its shell side but through a control valve with narrow pores to ensure huge expansion of this slip stream ammonia to vaporize resulting in Joule Thomson cooling effect to chill the main stream of ammonia in the tube side to a temperature of about (−)25 to (−)30°C which then flows to scrapper chiller series and solvent chillers as discussed earlier. The flashed ammonia vapor from this Joule Thomson chiller then flows to compressor suction vessel along with Joule Thompson scrapper chillers as mentioned earlier. For safety with respect to surging of the compressor, an anti-surge control is also provided in the compressor gas flow circuit and care should be taken for good operability of this anti-surge control circuit.

2.4.1.3.6 Vacuum Generation Circuit

We have discussed earlier that to ensure effective continuous filtration of wax cake from DWO-wax chilled slurry, filtration is to be carried out under vacuum; to establish the same, inert gas flow system should be established which should flow along with DWO-wax slurry from filter hood to filtrate pipes under the filter cloth to reach the filtrate vessel, and the inert gas vapor from top of the vessel should flow to the suction of a compressor which compress the inert gas marginally above atmospheric pressure, say, it compresses from 0.2 ata (atmospheric pressure absolute) at suction to 1.3 ata at discharge. Due to very high volume of inert gas handling for operating eight to ten numbers of rotary filters, the compressor selected is of multistage centrifugal nature and its discharge temperature rises to about 140°C which is then cooled in a water cooler to about 40°C followed by further cooling to subzero temperature, viz. to about (−)15°C through a refrigerant chiller using same ammonia vapor as discussed above to avoid any melting of the wax cake while the inert gas flows through the filter circuit. Like ammonia compressor, a spill back cum anti-surge control is also provided with inert gas compressor. The inert gas should be dry enough to avoid ice formation in the circuit thus preventing any choking in the flow path through the refrigerant chillers.

2.4.1.3.7 Filter Circuit

While the principles of need and operation of filters have been discussed earlier, additional points should be followed for efficient operation of rotary vacuum filters as follows.

To control the vacuum in filtration, filtrate is taken through three outlet lines of each filter instead of taking through a single header; the concept for this approach is that the filter, during its rotation, gets highest deposit / thickness of wax while moving through submergence of the filter vat; as the cake moves upward out of the submergence, due to filter rotation, its thickness gets reduced as no further wax cake formation but its entrapped oil is sucked out into filtrate pipes; with continuous rotation, when the cake reaches upward further to 90° position, its thickness gets further reduced due to spray of cold wash, on the filter cloth, which further washes away the entrapped oil in the wax into the filtrate pipes thereby further improving DWO yield as discussed earlier. Hence, adequate cold wash flow at filtration temperature should be ensured. To optimize the cold wash flow distribution, there are many spray nozzles pipes provided over the top periphery of the filter cloth, and it is the art of the operator to optimize the spray pattern.

In addition to control the filtrate flow and cold wash flow distribution as discussed above, filter rotation is another parameter to control the wax cake thickness and capacity of filtration. With increase of rotation, capacity of filtration is increased due to less buildup of cake thickness on the filter cloth which in turn increases the rate the filtration due to less pressure drop of the oil moving through reduced cake thickness. But if lower cake thickness is due to less formation of crystalline waxes (or due to formation of micro-crystalline waxes), then even with low thickness of wax cake, filtration rate would be poor due to less porosity of the filter cake which prevents building of further wax on the cloth thereby affecting filtration rate.

The doctor's knife kept on the surface of the filter cloth at the zone of wax blowing out from the filter cloth after its full rotation to a connected vertical vessel kept underneath the filter has been discussed earlier, but the point to be noted is that the filter cake can't be blown out as the filter is under vacuum. To facilitate such blowing of the wax cake, a small flow of inert gas led through a row of small pipes from filter cloth inside near the zone of complete 360° rotation of the filter and doctor's knife is just kept above that location without touching the filter cloth so that wax cake slurry moves down smoothly to the bottom vessel called filter boot.

Note: the schematic diagram of a filter is not shown in this chapter and the same is provided in Chapter 4 which is similar and can be used to understand the above-mentioned filtration process.

During continuous operation of the filters, some micro crystals of wax get blocked in the filter cloth pores thereby plugging the filtrate flow which in turn reduces the filtration capacity of the filters, i.e. feed processing capacity of the unit. To overcome this problem, such filters after a particular duration, say once in 8 hours are taken out of service and washed with hot solvent to melt out this wax and the filter become again efficient to take it in service. The washed solvent plus mixed hot wax can be routed to slop tank to mix it in the unit feed or can be routed to main filtrate stream depending upon the decision of the operation group to see how this small quantity of slop affect the main bulk DWO quality.

Filter cloth is wound on the filter drum with the help of steel winding wire; the end of the wire is tied up at one end which can be further tightened from outside. Due to time-to-time hot wash of the filters as discussed above, some elongation of the steel wires happens which unless tightened from time to time may cause wrapping of the wires leading to detachment of filter cloth from the filter drum resulting in

TABLE 2.13
Salient Case Study Operating Parameters of SDU

Sl. No.	Parameter	150 SN	500 SN	150 BS
1	Primary Solvent Ratio, V/V (S:F)	1:0.8	1:1	1:4
2	Secondary Solvent Ratio, V/V (S:F)	1:1	1:1	NIL
3	Tertiary Solvent Ratio, V/V (S:F)	NIL	0:0.5	NIL
4	Dewaxing Filter Temp., Deg.C.	(-)15	(-)15	(-)15
5	Spray Solvent Ratio in Filter, V/V	1:1.2	1:1.2	1:1.5
6	Solvent Spray Temp. in Filter, Deg.C.	(-)15	(-)15	(-)15
7	Filter Vat Level, % Max.	50	50	50
8	Filter Vacuum, Bar	(-)0.85	(-)0.85	(-)0.85
9	Filter Blanketing Pr. Bar abs.	1.1	1.1	1.1

TABLE 2.14
Desired Qualities of DWO, Slack Wax and Their Yields

Sl. No.	Attribute	150N	500N	150BS
1.	DWO:			
	-Viscosity, cst. @100°C	5.5–6.5	10–11	31–33
	-Sp. Gravity	0.855–0.875	0.875–0.895	0.895–0.915
	-CCR, %wt.	0.01–0.02	0.03-.04	0.8–0.9
	-Pour point, °C	(-)9 Max.	(-6) Max.	(-)3 to (-)6
	-Yield, %wt.	75	72	78
2.	Slack Wax:			
	-Sp. Gravity	0.85–0.87	0.875–0.895	0.89–0.90
	-Oil content, %wt., Max.	15	25	12

short circuit feed plus solvent to filtrate without filtration. Hence, maintenance group monitor the filters in this respect and takes care as mentioned above.

Note: except the last three points, all the operations as mentioned above can be mathematically demonstrated through equations 2.7, 2.8 and 2.9 as discussed earlier.

- **Case study operating conditions of SDU:**

Operating parameters to control dewaxing process are shown in Table 2.13.

- **Desired properties of DWO (dewaxed oil) and by-product, slack wax achieved in SDU:**

Maintaining the above operating conditions during processing of the raffinate of qualities as mentioned in Table 2.13 as shown earlier, the desired achievable qualities of DWO and slack wax are shown in Table 2.14.

- **Illustrations:**
  i. Q. DWO pour point reported off-spec., i.e. higher than normal. What should be corrective measure(s)?
     A. First to check that whether secondary dilution temperature and/or cold wash temperature has increased or not; if these are in normal values, then dewaxing temperature, i.e. final feed-solvent mix chiller outlet temperature should be reduced maintaining the temperature gradient in the chiller series; this would facilitate better crystallization of the wax to remain in solvent-wax phase and in turn would prevent passing of this wax into the DWO-solvent phase through the filter pores.

     Sometimes, with all operating parameters at normal values and at top of that doing further adjustment directionally as mentioned above don't bring down the pour point of DWO; in that case, the filters, one after other, should be taken out of service and to be inspected in situ to find out leak, if any, in the filter cloth; the filter with torn out cloth should then be kept out of service which in turns facilitate to get the desired pour point of DWO.
  ii. Q. Feed rate in SDU is restricted by increasing filter vat level to alarming level due to poor crystallization thereby causing poor separation between oil-solvent phase and wax-solvent phase which in turn reduces the filterability of the chilled slurry at the filters by clogging the filter cloth pores resulting in filter vat level increasing with poor rate of filtration. What remedial measure(s) should be taken?
     A. To overcome, first increase the rotation of the filters to maximum as per design which is generally 6 mpr (minutes per revolution) maximum; if the filters' vat level doesn't decrease, then chillers' outlet temperatures should be decreased by 1 or 2°C along with secondary dilution temperatures; even if the problem is not overcome, then first secondary dilution ratio should be decreased and as a next course of action, primary dilution ratio should be increased. As a last resort, feed rate should be decreased.

     Sometimes, if CCR of the feed is on the higher side than normal, the aromatic molecules being bigger in sizes may clog the filters' cloth thereby affecting the filtration rate and ultimately feed capacity; in that case, the quality of the feed (raffinate) should be improved in the upstream unit, i.e. aromatic extraction unit.
  iii. Q. DWO yield is lower than expected; what should be corrective measure?
     A. Secondary dilution ratio or temperature should be increased or dewaxing temperature to be marginally increased by 1 or 2°C or increase primary dilution ratio without affecting pour point of DWO or affecting filtration rate in the filters. Alternatively, cold wash flow rates can be increased which, however, marginally increase the DWO pour point without affecting filtration rate.

iv. Evaluation of loss in DWO recovery:

Q. Before taking the above action(s) to improve the DWO yield, first get analysis of by-product, slack wax with respect to its oil content; if oil content of slack wax is higher than the desired value, then the above action(s) should be undertaken to increase DWO yield.

For example, if slack wax oil content is higher at 28%wt. against normal value of 20%wt. max. with DWO yield achieved at 70%wt., how much is the scope to increase DWO yield?

A. With 100 MT feed, DWO yield is 70 MT, i.e. slack wax yield is 30 MT.

To improve slack wax oil content to 20%wt. from 28%wt. as mentioned above, oil to be recovered, say X MT, from slack wax into DWO phase is given by:

$$30 \times (28 - 20)/100 = 2.4 \text{ MT},$$

i.e. increase in possible DWO yield = 70 + 2.4 MT = 72.4 MT,
i.e. improved DWO yield = 72.4%wt. with reduction of slack wax yield to (30 –2.4) = 27.6 MT
i.e. 27.6%wt.

- **Improvement of DWO yield by dosing dewaxing aid in SDU operation:**
Dewaxing aid is a chemical, viz. alkylated phenol or chlorinated wax or polymethacrylate which when added to SDU feed at a very low dosing rate of about 100 to 500 ppm as established during trial operation, the problems as discussed in Illustrations (i) to (iii), can be solved without taking recourse to all tedious methods as mentioned in there.

But alkylated phenol and chlorinated wax, having their inherent problems affecting the processing equipment, wax color and above all, it affecting the downstream unit's operation and due to issues not approved by FDA (food and drug administration), USA, are not used in the industry, and instead, polymethacrylate of certain grades as approved by FDA, USA are used as dewaxing aid in SDU.

The dewaxing aid compound acts as a crystal modifier to the feed in SDU to help in nucleation of wax crystal by preventing the micro-crystalline nucleus, if any, to initiate the crystal growth. Even if proper nucleation gets started, the dewaxing aid can facilitate homogeneous crystal growth by getting adsorbed into the edges of nearby wax crystals; this in turn facilitates formation of platelet type or bigger needle type crystal agglomerates preventing the entrapment of oils inside the small wax crystals with crystal agglomerates thus creating good porosities in the wax cake thereby facilitating better filterability of the chilled slurry through the filters resulting in not only enhanced feed rate but also increase in DWO yield without resorting to all operational adjustments as mentioned above.

Polymethacrylate has no adverse effect in processing the product, DWO or by-product, slack wax in any subsequent catalytic or non-catalytic process unit.

- **Case study utility consumptions in SDU:**
Solvents like MEK and toluene are consumed due to losses from different leak sources if any and major loss to the blanketing system of inert gas.

TABLE 2.15
Case Study Utilities Consumptions in SDU

Utilities per MT of Unit Feed Capacity	Consumption Rate
MEK, Kg	1.5
Toluene, Kg	1.5
LP (70 psi) steam, Kg	250
MP (200 psi) steam, Kg	400
Electricity, KWH	130
Circulating Water, M^3	70

Note: the figures are for 500SN grade feed; for BS grade feed, utilities consumptions are 30% higher due to feed rate lower by 40% which can save only 10% utilities due to same unit is used for operation at lower feed rate while designed to operate at higher feed rate.

LP steam and MP steam are used in SDU; MP steam consumption becomes higher if no furnace is provided in DWO solvent circuit where solvent load is very high. Electrical power consumption is highest among all other process units in base oil manufacturing process. Typical average consumptions as a case study are given in Table 2.15.

- **Metallurgy and investment cost:**

Though this unit also involves separation process without chemical reaction, the operating cost is very high as understood from the utilities' consumptions figure from Table 2.12 because there are more numbers of sub-sections with high operating loads in each section. Accordingly, the number of pieces of equipment are also very high with some major capacity rotary equipment like compressors. But pressure of various sections remain in the range of two to five bar with exception in refrigeration circuit at about 15 bar and temperature also ranges from ambient to about 180°C with exception in refrigeration/chiller circuits where temperature remains at sub-zero; thus, carbon steel pipe and fittings of normal grades and with low carbon steel at sub-zero temperature are adequate for this unit.

With the above, approximate installed investment cost becomes 0.4 million USD per TMTA of unit feed capacity for a 300 TMTA feed capacity plant.

2.4.1.4 Hydro-finishing of Dewaxed Oil (DWO)

DWO obtained from SDU now have good properties with respect to VI as well as pour point while its basic viscosity was taken care of in VDU. Also, while improving its VI in FEU / NMPEU, viscosity change, i.e. viscosity reduction of raffinate, is also taken into consideration so that it does not go off spec and if necessary, distillate draw rate adjustment in VDU has to be done. Similarly, in SDU, while improving pour point of the raffinate to get desired quality DWO, there is negligible change in the viscosity, i.e. very minor increase in viscosity, but there would be some reduction of VI as waxy paraffins that are removed from raffinate have very high VI. Hence, while operating FEU / NMPEU, adequate cushion in VI value of the raffinate should be kept such that after dewaxing in SDU, final VI of the products does not go off-spec.

Even after taking care of all these as mentioned above, the DWO can't be fit to be used as LOBS for blending into lubricating oil formulation as mentioned earlier, i.e. there would remain the problem of dark color, color stability, and oxidation stability issues in DWO which are yet to be addressed. Hence, DWO is further processed in hydro-finishing unit (HFU) to take care of these properties.

- **Principles in HFU operation:**

HFU is a reaction process unit and not a physical separation process like CDU, VDU, FEU / NMPEU and SDU. The dark color in DWO is due to higher presence of impurities like nitrogen. The removal or reduction of nitrogen content not only improve the color but improves the color stability. Similarly, presence of very high concentration of sulfur causes the corrosion issues to the machines where the lubricant to be used, but presence sulfur contributes to good oxidation stability property in the lubricant as discussed earlier. Hence, in HFU, there is no need to remove sulfur completely but only to reduce its concentration to some extent. Also, there may be some unsaturates in the DWO though very rare except in the side chains of some aromatic compounds that are also of much lower concentration in most of lube bearing crude oils.

- **Chemical reactions:**

In HFU, hydrogen is used to convert nitrogen into NH_3 (ammonia), part sulfur is converted into H_2S (hydrogen sulfide), olefins, if any, are saturated into paraffins.

The reactions are liquid-vapor mix phase reactions, i.e. feed is in the liquid form and hydrogen is in the gas form and the reactions occur in a fixed bed of catalyst. In this liquid-vapor phase reaction, hydrogen gas gets absorbed into the liquid pool of feed; the liquid feed gets adsorbed into the active pores of catalyst surfaces and the hydrogen absorbed into the feed liquid then gets similarly adsorbed into the catalyst active pores thereby raising reaction energy level facilitating the desired reaction to take place and the reaction product(s) get desorbed from the catalyst surfaces to exit along with reactor effluent.

The reactions stoichiometries are given as follows:

Denitrification:

$$NH_2 + H_2 \rightarrow +2NH_3 \quad (2.12)$$

Desulfurization:

$$2R\text{-}S + 3H_2 = 2R\text{-}H + 2H_2S \quad (2.13)$$

Where R is alkyl or phenyl group in hydrocarbon series.

Olefin saturation:

$$R\text{-}CH\text{=}CH\text{-}R + H_2 = R\text{-}CH_2\text{-}R\text{-}CH_2 \quad (2.14)$$

It is seen from the above that no hydrogenation occurs inside the closed ring structure of the hydrocarbons, that's why it is called mild hydrotreating unit or hydrofinishing unit.

With the heavier feed like DWO, the pressure required for the reactions to occur is as high as 65 Kg/cm^2g and reaction temperature is maintained from about 230°C to 300°C depending upon the grade of DWO used.

- **Reaction mechanism:**

Reactions, viz. Equations 2.12 and 2.13 are the main reactions and reaction shown by Equation 2.14 is not predominant as there is negligible presence of olefin in the feedstock obtained from crude source and no other feed obtained from cracking unit is added to the feed.

The reaction in Equation 2.13 is called desulfurization which follows first order kinetics and reaction in Equation 2.12 is called denitrification which follows second order kinetics.

As in the group-I LOBS, because there is no severe constraint on sulfur content, viz. 0.5%wt., 1.0%wt., 1.5%wt., and 2.0%wt. max. for 100SN, 150SN, 500SN, and 150BS grades respectively as seen in Table 2.5, it is not necessary to remove all sulfur from the feed, rather it is better to have some sulfur in the product as sulfur acts as an anti-oxidant preventing degradation in storage.

Also, the presence of nitrogen contributes to the color of product; hence, the objective should be to remove nitrogen as much possible to get the light color in the product; the specifications on products' color restrict the nitrogen content in the products to 100ppm, 150ppm, 200ppm max. for 100SN, 150SN, and 500SN grades respectively in order to get the improved color in the products; but for 150BS grade, the requirement is only to report the nitrogen content of the product and the same is indirectly restricted by ASTM color specification of the products as seen from Table 2.5.

So, in HFU, more thrust is provided in the nitrogen removal from the feed than sulfur removal as discussed above.

The above reactions of hydrogenation are facilitated by using the right catalyst, and hence, selection of the catalyst and maintaining its activity throughout the operation cycle is very important for energy savings as well as for achieving product qualities on sustained basis till the turnaround shutdown of the unit.

In addition to the right selection of the catalyst, hydrogen purity along with its flow, reaction pressure, temperature, and residence time of the feed, i.e. space velocity are also important to ensure satisfactory level of performance in the reactions as further explained below.

- Hydrogen purity, flows and system pressure:
Higher hydrogen purity in hydrogen make-up and recycle gas together increases the mole fraction of hydrogen and higher mole fraction along with higher system pressure (facilitated by higher make-up hydrogen gas flow) increase the partial pressure of hydrogen for reaction which in turn dictates the kinetic rate of the reactions. The partial pressure is expressed mathematically as below:
Hydrogen purity × reactor pressure = partial pressure of hydrogen.

- Reaction temperature:
In the kinetic equations of any order (first order, second order, etc.), the parameter, temperature is there in the reaction rate constant established by Arrhenius equation of rate constant, which shows increase in the temperature, increases the rate of the reaction.
- Residence time / space velocity:
Residence time is the time in hour or minute which the feed takes to traverse the full bed of the catalyst from inlet to outlet. The inverse of this time is called reaction space velocity which is generally expressed as 'hr^{-1}'.
More residence time, i.e. less space velocity means higher rate of the reaction.

Note: it is preferred that feed flow occurs from reactor bottom to top to ensure better mixing of the feed with catalyst and hydrogen, but in case where there is higher feed rate as compared to quantity of catalyst required, there is fluidization of the catalyst due to higher velocity of the feed thereby damaging the catalyst bed by carrying away the catalyst by breaking its retainer plate. In HFU, the feed enters the reactor from the top and leaves from the bottom based on the above analysis.
- Catalyst activity:
Higher catalyst activity means reaction(s) can be achieved at a lower reaction temperature thus saving energy consumption along with avoiding the high temperature constraints like thermal cracking of feed at higher temperature and reactor-furnace system design limitations. The catalyst activity decreases during operation in the long run due to coke deposition in the micro pores of the catalyst by undesired thermal cracking in addition to desired hydrogenation reactions when catalyst regeneration/replacement should be done. In addition to evaluate catalyst activity through performance analysis, catalyst sampling devise is also provided at reactor bottom for on-line sampling and laboratory analysis.
- **Selection of the catalyst:**
For hydrogenation reactions, Co-Mo (cobalt-molybdenum) or Ni-Mo (nickel-molybdenum) based catalyst, i.e. either of these two combination metals are impregnated on alumina base (Al_2O_3) and the catalysts are produced in cylindrical shape extrudate form to be used as catalyst bed. Co-Mo catalyst can perform high degree of de-nitrification as compared to Ni-Mo catalyst whereas de-sulfurization performance of both these catalysts are comparable. When aromatic extraction unit is used in LOBS making process, it means major amount of nitrogen is already removed along with the aromatics. Hence, in HFU, Ni-Mo catalyst is used instead of Co-Mo catalyst thereby saving catalyst cost.

We know that sulfide form has more surface area than oxide, i.e. NiS-MoS catalyst would have more surface area than NiO-MiO thereby providing more active surface for the reactions. But handling of the sulfide form is unsafe as compared to oxide form; hence, catalyst is procured from distant supplier in oxide form and then it is converted into sulfide form inside the reactor through hydrogenation reaction with hydrogen supplied by liquid DMDS (di-methyl di-sulfide) at a different set of operating conditions like lower pressure of about 30–40 bar and temperature of about 230°C for completion of bulk of the reactions and finishing it for a small duration at a higher temperature of about 300°C. The reactions are shown as below:

$$2H_2 + CH_3\text{-}S\text{-}S\text{-}CH_3 = 2H_2S + C_2H_6 \tag{2.15}$$

$$NiO\text{-}MiO + 2H_2S = NiS\text{-}MiS + 2H_2O \tag{2.16}$$

Note: from above, it is evident that catalyst should always remain in sulfide form during feed processing, i.e. its sulfur should not be removed (called leached out) by reaction with hydrogen while the sulfur from the feed is removed by reaction with hydrogen; to prevent this, there should be minimum presence of H_2S in the recycle gas, viz. 50 to 100 ppm. When it goes down to below 50 ppm (as measurement of recycle gas done daily by laboratory), DMDS injection should be started to get back the recycle gas desired H_2S within 50–100 ppm. However, higher H_2S concentration above this level would act adversely by reducing the desired reaction rate of hydrogenation of feed.

- **Process description of HFU – a case study:**

DWO is taken from the storage tank at a temperature of about 70°C to HFU where the feed is pumped to a discharge pressure of about 70 Kg/cm^{2}g to flow through a heat exchanger to exchange heat with the hot product coming out from the product steam stripping / drying column followed by series of heat exchangers to exchange heat with the hot fluid coming out from the hydro-finishing reactor. Hydrogen is supplied for this reaction process in two ways, viz. make up hydrogen for the reaction is supplied by a compressor at a pressure of about 70 Kg/cm^{2}g and joins the feed pump discharge where recycle hydrogen is also supplied by a recycle hydrogen gas compressor; recycle gas compressor is used to provide minimum hydrogen / hydrocarbon mole ratio to ensure adequate hydrogen partial pressure in the reactor in order to avoid any thermal cracking/coking of the hot feed. The recycle hydrogen gas is supplied by a separate gas compressor at the same pressure as mentioned above to join the feed and make-up hydrogen gas streams. The mixed feed and gas then flow through a series of heat exchangers as mentioned above followed by heating through a furnace before entering a fixed bed reactor. The mixed stream enters a vertical fixed bed reactor from its top at a temperature of about 230°C to 300°C depending upon the grade of DWO used. Reaction pressure becomes about 65 Kg/cm^{2}C due to pressure drop in those heat exchangers. The catalyst used in the reactor is Ni-Mo extrudates as mentioned above. There is very low exothermicity in all those hydrogenation reactions due to much less conversion in the reactions as the concentration of olefins which causes most exothermicity is negligible, also the degree of desulfurization and denitrification is too low to increase the temperature of whole mass of DWO to a measurable rise in temperature and that's why many a time there is no temperature rise observed in the reactor.

The reactor outlet effluent leaving from reactor bottom then enters the heat exchangers as mentioned earlier where it is cooled to about 120°C to 160°C depending upon grade of DWO used. The cooled mixed stream then enters a flash separator vessel at same high pressure of about 65 Kg/cm^{2}g where the hot hydrocarbon vapors like unreacted hydrogen and some light hydrocarbon vapors exit from the top of that vessel followed by further cooling in water cooler to send it low temperature flash vessel from where the gas comes out at a temperature of about 45°C which then is used as recycle gas as mentioned above but via amine wash circuit to absorb the H_2S in the recycle gas.

The bottom of the high temperature flash vessel as mentioned above is routed to a low-pressure flash vessel without further cooling. The bottom of the low-pressure flash vessel flows to a steam stripping column via a steam heater to heat it up to about 180°C and the bottom of the stripper moves by gravity to a vacuum stripper to remove the lighter hydrocarbon to take care of the flash point of finished LOBS. The vapor from both of the stripper and vacuum drying column is condensed in an overhead condenser followed by routing the condensed hydrocarbon to a vessel from where it is pumped to slop tank. The top vapor from high temperature low pressure flash vessel is routed to a water cooler followed by routing it to a vessel called low temperature low pressure flash vessel from top of which the gas is routed to fuel gas system of the refinery via a low-pressure amine wash unit to remove H_2S from it.

In the amine wash circuit inside HFU, the sour recycle gas enters an absorber column from the bottom where liquid amine, viz. DEA (di ethanol amine) enters from the top. The absorption temperature is ambient while the pressure is about 65 Kg/cm^{2}g in order to be in parity with recycle gas pressure. To push DEA at such high pressure, generally a reciprocating pump is used as the flow requirement is very low at about 4 to 5 M^3/ Hr. for DWO feed rate of about 50 M^3/ Hr. with feed sulfur content of about 2%wt.

The bottom of the vacuum drying column is heat exchanged with feed as mentioned earlier and/or with stripper inlet feed as mentioned above depending upon heat exchanger circuit design in specific cases. The product stream after this heat exchanger is further cooled in a water cooler and pumped to finished LOBS storage tank.

The case study schematic flow diagram for the process depicting stream flows, process control and heat exchangers network is shown in Figure 2.8.

- **Case study operating conditions of HFU:**

Case study operating parameters to control the main reaction process are given in Table 2.16.

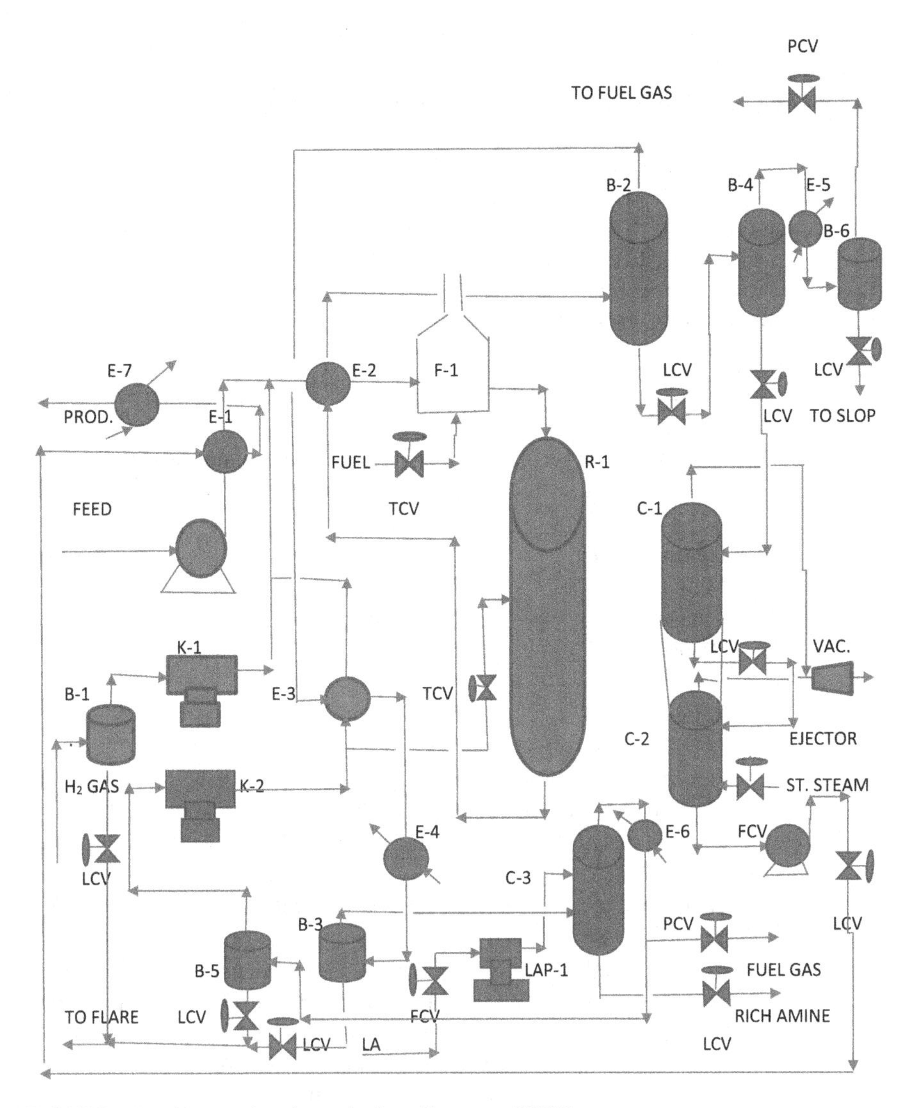

FIGURE 2.8 Case study schematic flow diagram of HFU.

- **Desired product qualities and yield:**

To meet the product specifications as shown in Table 2.5 is the ultimate objective to be achieved after processing DWO in HFU to get the finished LOBS.

As the catalytic reactions in HFU are mild ones, yield of the product is nearly the same as the feed rate, viz. yield of about 99%wt. is achieved in HFU.

- **Illustrations:**
  i. Q. In HFU, feed rate is 50 m^3/hr., hydrogen gas make-up flow rate is 150 Kg/hr., gas composition in volume: H_2-78%, CH_4-12%, C_2H_6-8%, C_3H_8-2%. Find out make-up gas ratio in Nm^3/m^3.

TABLE 2.16
Case Study Operating Conditions in HFU for Various Grades

Sl. No.	Parameter	150 SN	500 SN	150 BS
1	Make-Up (MU) H_2 Gas to Feed ratio, Nm^3/m^3	10:1	10:1	10:1
2	Recycle Gas (RG) to Feed ratio, Nm^3/m^3	100:1	100:1	100:1
3	Furnace outlet/Reactor Temperature, °C	210	230	300
4	Reactor Pressure, Bar	65	65	65
5	Reaction space velocity, Hr^{-1}	1.6	1.9	2.0
6	MU Gas H_2 purity, %wt.	70–80	70–80	70–80
7	RG H_2 purity, %wt.	70–75	70–75	70–75

A. The molecular weight the make-up gas is calculated as below:

Component	wt.
Hydrogen (H_2)	78 × 2 = 156
Methane (CH_4)	12 × 16 = 192
Ethane (C_2H_6)	8 × 30 = 240
Propane (C_3H_8)	2 × 44 = 88

Total = 676 in 100 moles of make-up gas.

Hence, molecular wt. of make-up gas = 676/100 = 6.76
Make-up gas flow rate in kg-moles/hr. = 150/6.76 = 22.2
So, volumetric flow rate of the gas = 22.2 × 22.4 Nm^3/hr = 497.3 Nm^3/hr.
Hence, make-up gas to feed ratio = 497.3/50 Nm^3/m^3 = 9.94:1 Nm^3/m^3.

ii. Q. In HFU, a catalyst bed of 17 MT with catalyst density of 0.655 is used for hydro finishing of DWO at a feed rate of 50 M^3/Hr. what is the space velocity of the reaction?
A. The volume of the catalyst = 17/0.655 = 25.955 M^3.
So, the reaction space velocity = 50/25.955 = 1.93 Hr^{-1}

iii. Q. The yield in HFU, SDU, FEU are 99%wt., 75%wt., and 65%wt. respectively of all the lube bearing distillates of vacuum distillation unit; also, yield of DAO is 25%wt. with its similar yield at FEU, SDU and HFU as mentioned above. The vacuum feedstock is 48%wt of crude oil. The total lube bearing distillate yield in vacuum distillation unit is 50%wt. and vacuum residue yield is 45%wt. what is the lube potential of the crude oil?
A. DAO potential = 45 × 0.25 = 11.25%wt.
Hence, overall lube distillate in VDU = 50 + 11.25 = 61.25%wt.
Hence, the Lube potential of the crude oil
= 61.25 × 0.99 × 0.75 × 0.65 × 0.48 = 14.2%wt.

TABLE 2.17
Case Study Utility Consumptions in HFU

Utilities per MT of Unit Feed Capacity	Consumption Rate
Hydrogen, Kg	1
LP (70 psi) steam, Kg	25
MP (200 psi) steam, Kg	45
Fuel, Kg	10
Electricity, KWH	40
Circulating Water, M^3	16

Note: the figures are for 500SN grade feed; for BS grade feed, utilities consumptions are 30% higher due to feed rate lower by 40% which can save only 10% utilities due to same unit is used for operation at lower feed rate while designed to operate at higher feed rate.

- **Case study utilities consumptions in HFU:**

It is reaction unit unlike physical separation units as described earlier above. In the reaction hydrogen is required as it is a hydrotreating unit, but consumption of hydrogen is low as it is a mild hydrotreating unit; however, it requires high pressure reaction section of about 60 bar.

The utility consumptions are given Table 2.17.

- **Metallurgy and Investment cost of HFU:**

As mentioned above, the unit being a reaction unit requires a high-pressure reactor of about 60 bar pressure. The unit contains three sub-sections with two sub-sections at high pressure of about 60 bar, i.e. first one is reaction section including reactor, two hydrogen compressors, one high pressure feed pump along with piping and exchangers, second one high pressure gas sweetening section containing amine wash column and associated high pressure pump; the third section is low pressure section of about 5 bar pressure up to a vacuum column for product drying associated with piping, heat exchangers and pumps. Due to having reaction sections at high pressure of about 60 bar and at a temperature of 220 to 300°C, Alloy steel (austenitic stainless steel) in high pressure reactor, heat exchangers, furnaces are required to be used. However, the unit is a compact unit with minimum equipment unlike physical separation processes as described earlier above.

With the above, the approximate installed investment cost becomes 0.4 million USD per TMTA of unit feed capacity for 250 TMTA plant capacity.

2.4.2 Processes for Group-II and Group-III LOBS Production

As the Group-I LOBS have limitation in pour point and VI as seen in the Section 2.2, the Group-II and III LOBS have come into place to meet the requirement of machines come up with advancement in technologies requiring high speed of operation and to be efficiently adapted in any cold climate. Hence, quality improvement in lubes becomes evident with respect to VI and pour point mainly. But we have understood the limitation of physical separation processes from discussion on Group-I LOBS to achieve the higher VI requirement and lower pour point requirement. To overcome, catalytic dewaxing technology has come into place where the first objective is to saturate the aromatics or any heterocyclic compound into naphthenes, and saturate the olefins into paraffins followed by isomerization of the normal paraffins in the waxes of the hydrotreated product into non-waxy iso-paraffins thereby increasing the value of VI and lowering of pour point respectively. The new grade of LOBS produced is called Group-II LOBS with maximum VI achievable up to 120. If the process severity is suitably increased or if the feedstock quality is so better that contains less aromatics, that product LOBS of VI more than 120 can be achieved, then the LOBS produced can be called Group-III LOBS.

It appears from above that yield of lube/LOBS would be very higher as the aromatics instead of physically separated are converted into lube /LOBS, but that would not be the case as during conversion process there would be some undesired cracking to generate some non-lube compounds like diesel, naphtha, LPG and gas thereby restricting the yield of lube. Also, both waxy and non-waxy paraffins would get cracked into such byproducts as mentioned above. However, these byproducts fetch more value than the byproducts, viz. oily aromatics (called aromatic extract) and oily wax (called slack wax) produced in physical separation process as discussed earlier. Another advantage of catalytic dewaxing is that as the process is carried in presence of hydrogen, color and color stability problems are automatically addressed and thus there is no requirement of processing the finished product through a separate hydro finishing unit (HFU). But yield of LOBS in catalytic dewaxing becomes less than 70%wt as compared to 72 to 75%wt achieved in physical separation process due to cracking of paraffins. To overcome this limitation, a second stage reaction is followed in the present days of catalytic dewaxing unit like isomerization where instead of cracking, the waxy paraffins get isomerized into branch chain compounds thereby improving pour point, i.e. achieving lower pour point of LOBS at the same time increasing LOBS yield. With this second stage reaction, yield of LOBS in catalytic dewaxing unit, the yield of LOBS improves from 70% and below to 75 to 80%wt.

Catalytic dewaxing is carried out at a very high pressure of about 100 to 130 bar and at a temperature of about 300°C to achieve hydrogenation of the closed ring structures in aromatics and heterocyclic compounds (including denitrification and desulfurization of heterocyclic ring structures) in addition to saturation of open chains (including desulfurization and denitrification) which can be achieved at lower pressure. Due to this extensive desulfurization and denitrification, the color of the finished product, LOBS becomes water white including that of byproduct, diesel also. But one disadvantage due to extensive desulfurization is that oxidation stability of LOBS gets affected as discussed earlier. To take care of this problem, generally some external chemical like anti-oxidant is dosed in the finished product line before sending it to storage tank. Catalytic dewaxing can be carried out starting with vacuum distillate instead of raffinate from FEU as feed as understood from above discussion. But if physical separation of vacuum distillate is done in FEU or in NMPEU to take raffinate as feedstock of catalytic dewaxing unit instead of directly feeding of vacuum distillate to catalytic dewaxing unit, then the severity of catalytic dewaxing gets reduced and the reaction can be carried out at lower pressure of about 100 bar. Also, due to operation at lower severity, undesired cracking into fuel components can be reduced thereby increasing LOBS yield to about 75 to 80%wt.

Due to isomerization reaction occurring in the unit the catalytic dewaxing unit is also called a catalytic iso-dewaxing (CIDW) unit. One limitation on following isomerization is that in this process some undesired reactions like aromatization of some naphthenes occurs thereby again reducing VI of the LOBS to some extent. To overcome this problem, a third reactor is provided where outlet products from iso-dewaxing unit moves to a reactor where again mild hydrogenation is carried out to covert these aromatics back to naphthenes without causing much additional undesired cracking to keep LOBS yield at optimum level.

Earlier when catalytic dewaxing wax commercialized, it did not gain popularity due to not availability of isomerization technology to get optimum yield of LOBS. MOBIL, USA was first to commercialize catalytic dewaxing without isomerization. Subsequently, with the implementation of iso-dewaxing technology, both Chevron, USA and EXXON-MOBIL, USA have commercialized the production of Group-II LOBS on sustainable success basis particularly adding de-aromatization reactor to this process as discussed above.

The catalytic iso-dewaxing (CIDW) process as followed now-a-days is described in detail with all principles of the process as follows.

2.4.2.1 Catalytic Iso-dewaxing of Vacuum Distillate / Raffinate of FEU or NMPEU

In catalytic iso-dewaxing (CIDW) unit, basically three reaction sections are there as discussed above, viz. mild/moderate hydrocracking / hydrotreating followed by iso-dewaxing followed by de-aromatization section. To understand the process, it is necessary to understand the principles of these three reactions.

The principles and sequences of the reaction processes in hydrocracking / hydrotreating is explained as follows.

- **Principles of Hydrocracking / Hydrotreating:**

Regarding kinetics, hydrocracking is called HDS + HDT + HCU, i.e. hydrocracking reactions are mostly in series with some parallel reactions as shown below as group of equations, Equation 2.17 as below:

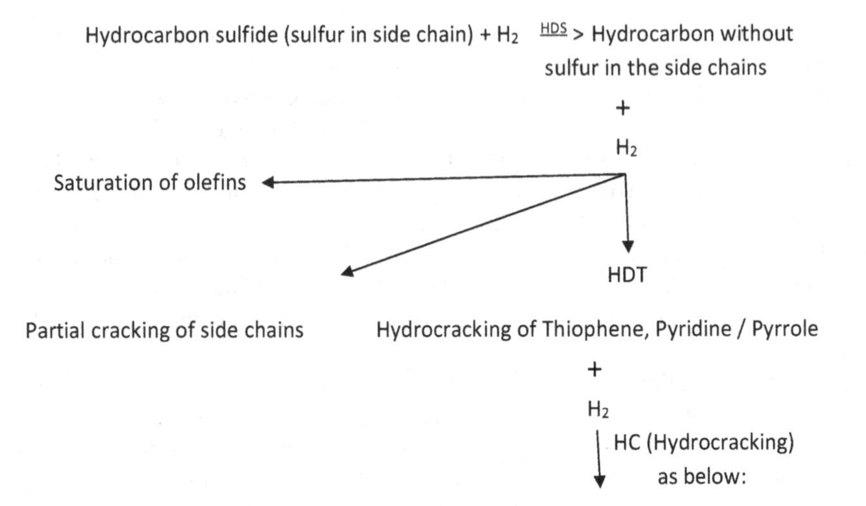

Note:

1) HDS, Olefin saturation, HDT, HC follow first order kinetics; only denitrification follows second order kinetics.
2) For hydrocracking of heavy gas oil feed mix to produce fuel oil instead of LOBS, operating pressure, temp. Hydrogen purity and H_2/HC ratio in hydrocracking are about 180 bar, 390°C, 96% and 300 Kg. moles/(M^3/Hr.) respectively. Approx. H_2 consumption is 3% wt. of feed rate.

For hydrocracking of vacuum distillate with the objective to LOBS, i.e. full hydrocracking is not required, and instead hydrotreating (HDT) of vacuum distillate as shown above is adequate, hence reaction pressure of about 130 bar and temperature of about 300°C found to be adequate.

If the Lube refinery has already a FEU or NMPEU for production of Group-I base oils, and the refinery desires to upgrade base oils quality from Group-I to Group-II, they should better take the raffinate from FEU or NMPEU as feed to lube hydrotreater. Also, in that case, as aromatic content of raffinate is less than 25%wt. as compared to about 40%wt. in the vacuum distillate, the severity of hydrotreating would be lower than using vacuum distillate as feed to hydrotreating, i.e. a reaction pressure of about 100 bar should be enough. This hydrotreating is called medium hydrotreating and advantage is that yield of the hydrotreated product would be higher than the case using vacuum distillate as feed for severe hydrotreating as mentioned above due to avoiding undesired cracking in the former case.

- **Catalyst used in hydrotreating:**
  Different types of catalyst are used in different beds of the reactor. In the first bed, some de-metallization catalyst should be placed to trap some metal

contaminants to ensure better performance of hydrotreating catalyst; then desulfurization / denitrification catalyst like cobalt-molybdenum impregnated silica-alumina catalyst is used in the second bed, and in the last bed specially manufactured zeolite catalyst is used to facilitate hydrotreating reactions. Zeolite is also silica alumina catalyst but crystalline in shape as manufactured to get desired reaction kinetics.

- **Hydrotreating reaction constraints:**

  Space velocity of about 1.0 is found to be optimum with raffinate as feedstock, while it may have to be kept lower with vacuum distillate as feedstock. But in any case, limiting space velocity in lube hydrotreating is 0.5.

  H_2 / HC recycle ratio of about 400 is optimum to achieve the reactions as well as to prevent undesired thermal cracking.

  Cold H_2 partial pressure at reactor is found to be optimum at about 120 bars for vacuum distillate dewaxing and it is lower for raffinate dewaxing.

  Reaction temperature varies with the grade of the feedstock used, and it is found to be optimal between 300°C and 320°C with feedstock ranging from 100N to 150BS grade equivalent. However, these temperatures are at start of run (SOR) conditions and with operation run length for about two years, the end of run (EOR) temperature increase by about 20°C.

  Maximum weighted average bed temperature (WABT) of about 380°C is considered for lube hydrotreating with the catalysts commercialized so far.

  There is exothermicity of about 10°C as observed in lube hydrotreating reaction.

  Impurity limitation in the reaction gas generally should be 2400 ppmw maximum for H_2S, and for nitrogen, it should be within 140 ppmw.

- **Principles of Iso-Dewaxing:**

The hydrotreated raffinate or distillate from the lube hydrotreater as mentioned above is fed to iso-dewaxing reactor where the objective is to isomerize the waxy paraffins into iso-paraffins instead of cracking of these paraffins. The reaction stoichiometry is shown as Equation 2.18 as follows:

$$H\text{-}(CH_2\text{-}CH_2\text{-}CH_2)_n\text{-}R \longrightarrow H(\overset{CH_3}{\overset{|}{C}}\text{-}CH_2)_n\text{-}R \quad (2.18)$$

The isomerization reaction occurs at marginally lower temperature than hydrotreating. In the above equation, the straight chain paraffin having higher VI but pour point also on the higher side whereas after the reaction, the branched chain paraffin, i.e. isomer of the feed paraffin has very low pour point as compared to feed whereas VI deteriorate very marginally with viscosity increase also marginally.

- **Catalyst used in iso-dewaxing and reaction isotherm:**

  For this reaction also zeolite catalyst is used but the same is manufactured in such conditions and with know-how that the crystals of the zeolites would have adequate pores size so that paraffins don't get trapped inside the pores and instead pass through the pores and during the passage of n-paraffins through the pores, isomerization reaction would occur. If the pores sizes are not made properly and paraffins get trapped in it, then cracking of the paraffins would occur like normal hydrocracking and / or hydrotreating.

In isomerization reaction kinetics, both metal function and acidic function of the catalyst are required because in this reaction kinetics though not shown in detail in the above Equation 2.18, it is worth mentioning that as a first step, normal paraffins gets converted into olefins by the metal function of the catalyst; this reaction is also called dehydrogenation. In the second step, the olefins so formed would get converted into iso-paraffins by the acid functions of the catalyst; this isomerization step is also called hydrogenation. Zeolites generally from alumina-silica have acidic properties but to have metallic properties certain metal(s) are impregnated during manufacture of such catalyst.

- **Reaction constraints in wax isomerization / iso-dewaxing:**
  At par with hydrotreating reaction, here also, a space velocity of about 1.0 is found to be optimum with raffinate as feedstock, while it may have to be kept lower with vacuum distillate as feedstock, and limiting space velocity is 0.5 minimum.

  H_2 / HC recycle ratio at marginally lower than 400 due to generation of impurities like H_2S, NH_3 and some light hydrocarbons in the hydrotreating reactor is found acceptable in the iso-dewaxing reactor.

  Cold H_2 partial pressure at reactor is found to be optimum at about 110 bar to facilitate keeping this reactor in series with hydrotreater reactor in flow dynamics.

  Reaction temperature varies with the grade of the feedstock used, and temperature at marginally lower by about 5°C than in hydrotreating reactor found to be adequate for iso-dewaxing reaction.

  Maximum weighted average bed temperature (WABT) of about 380°C is considered for lube hydrotreating with the catalysts commercialized so far.

  Though isomerization reaction is nearly thermo-neutral with lighter feedstock like naphtha, but with lube feedstock, the isomerization of if waxy paraffins found to be exothermic like hydrotreating reaction as mentioned above and an exotherm of about 14°C in the first bed narrowed down to about 10°C in the third bed is observed.

- **Principles of De-Aromatization:**
  As discussed earlier, in the iso-dewaxing process, some undesired aromatization reaction occurs even with optimized formulation / manufacture of iso-dewaxing catalyst. This in turn partly reduces the VI improvement achieved in the hydrotreating process earlier. To overcome this problem, the iso-dewaxed product from the isomerization process is further processed in a third reactor in presence of aromatic saturation catalyst to carry out mild saturation process. The reaction is shown as Equation 2.19.

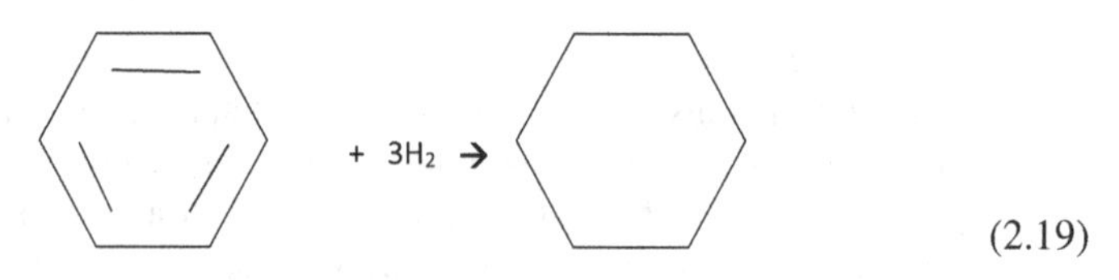

(2.19)

The above de-aromatization reaction is carried out at much lower temperature than in hydrotreating discussed earlier as only de-aromatization reaction is required to happen. The catalyst used in this reaction is noble metal catalyst and is called an aro-sat catalyst.

- **Reaction constraints in de-aromatization reactor:**
  As explained above, reaction temperature here is less by about 60 to 70°C than in hydrotreating and iso-dewaxing reactions. The reaction can be achieved keeping the reactor in series after the iso-dewaxing reactor thereby having a marginal pressure drop in flow dynamics to get a reaction pressure near to the iso-dewaxing, and H_2/HC ratio is also marginally less than in the first and second reactors (due to the generation of impurities as discussed above) which is acceptable.

- **Process description of catalytic iso-dewaxing (CIDW) unit – a case study:**

Vacuum distillate or raffinate from FEU / NMPEU as feed is taken from storage tank to the unit where it is pumped at a high pressure of about 115 bar to a series of heat exchangers to exchange heat with hot reactor effluent stream coming from aro-sat reactor and hydrotreater reactor. Make up and recycle hydrogen gas stream also join the feed before the heat exchanger train with the discharge pressure of the respective compressors at about 115 bar. The feed gas mixture exiting from the heat exchangers train enters the furnace to heat it up to about 300°C before it is routed to the hydrotreater reactor from its top. The reactor is vertically mounted with fixed bed catalyst but with multiple numbers. of catalyst bed. As the reaction is exothermic, to achieve the desired conversion, quench gas flow is maintained after each bed of catalyst. A slip stream of recycle hydrogen gas exiting from compressor discharge is cooled through a cooler and then injected to reactor as quench gas as mentioned above. To achieve higher efficiency of conversion, specially designed feed-gas mixture distributor is installed at the reactor top as well as at the entry of each catalyst bed to prevent channeling of the liquid-vapor mixture throughout the reactor. The reactor effluent exiting from reactor bottom marginally cooled before entering the second reactor, viz. iso-dewaxing (ISDW) reactor from its top along with hydrogen gas stream as reaction temperature of ISDW is about 5°C lower (at about 295°C) than in hydrotreating reaction. The ISDW reactor is also vertically mounted fixed bed reactor with multiple catalyst beds with gas quenching facility like in hydrotreating reactor as mentioned above. The reactor effluent from the bottom of ISDW reactor is further cooled to about 230°C in the heat exchanger train before it enters third reactor, viz. aro-sat reactor from its top. This reactor is also vertically mounted fixed bed reactor but with only two beds but here reaction is exothermic though marginal. The reactor effluent exiting from reactor bottom is cooled in the heat exchangers train as mentioned above and then moves to flash separator vessel where the unreacted hydrogen gas along with lighter hydrocarbon formed exit from the top of the vessel. This gas mixture is then further cooled to about ambient temperature followed by processing in Amine wash column to remove H_2S and NH_3 from this sour gas before it is routed to recycle gas compressor suction. The liquid effluent from bottom of the vessel is sent to a low-pressure flash vessel through a throttling control valve where further flash out vapor takes place, but this vapor exiting from vessel top is sent to refinery fuel gas network after passing through water cooler followed by passing through a low pressure amine wash column. The hot effluent leaving the low-pressure flash vessel bottom is then routed product recovery section, i.e. it is routed to a flash distillation column from where byproduct like diesel only comes out as a side cut of the column if unit feed is of 150N category. LPG and Naphtha exit as overhead vapor of the column which is then cooled and sent to a stabilizer distillation column to recover LPG from its top and naphtha from its bottom. The bottom of the flash distillation column is the main product, viz. LOBS. If unit feed is of 500N category, then byproduct LOBS, viz. 150N also comes out as side cut of the flash distillation column in addition to diesel as another side cut.

Similarly, if unit feed is of 150BS category, then one can get side cuts as diesel, 150N and 500N but in actual in the industry, provision is not made to draw so many side cuts, and as per specific design, it may be that with 150BS category of feed, side cuts are diesel and 500N category only with bottom of the column as the main 150BS LOBS.

The products as usual are cooled in heat exchanger network and further cooled in respective water coolers before sending these to respective storage tanks. As discussed earlier, anti-oxidant property of LOBS comes down after processing in this unit, and hence, anti-oxidant like DBPC is dosed in product run-down line to get back the oxidation stability property of the LOBS.

The case study schematic flow diagram of CIDW unit depicting stream flows, process control and heat exchangers network is shown in Figure 2.9.

- **Case study yield pattern in CIDW Unit:**

With Middle East crude source, the yield patterns are optimized as shown in Table 2.18.

- **Case study operating parameters:**

To achieve the above yield pattern, main operating parameters are given in Table 2.19.

- **Desired product qualities:**
  The desired product qualities achieved in this unit meet the specifications as shown in Table 2.4 earlier.

- **Case study utilities consumptions in CIDW unit:**

It is also a reaction unit like HFU as discussed earlier but more complicated than HFU as there are multiple reactors here that too with higher severity of operation. In the reactions, though hydrogen flow is required in all the three reactors but its main consumption is in the first reactor. Due to reactions at very high temperature like about 300 to 350°C, fuel is consumed to heat up to reaction temperature. The utility consumptions are given in Table 2.20.

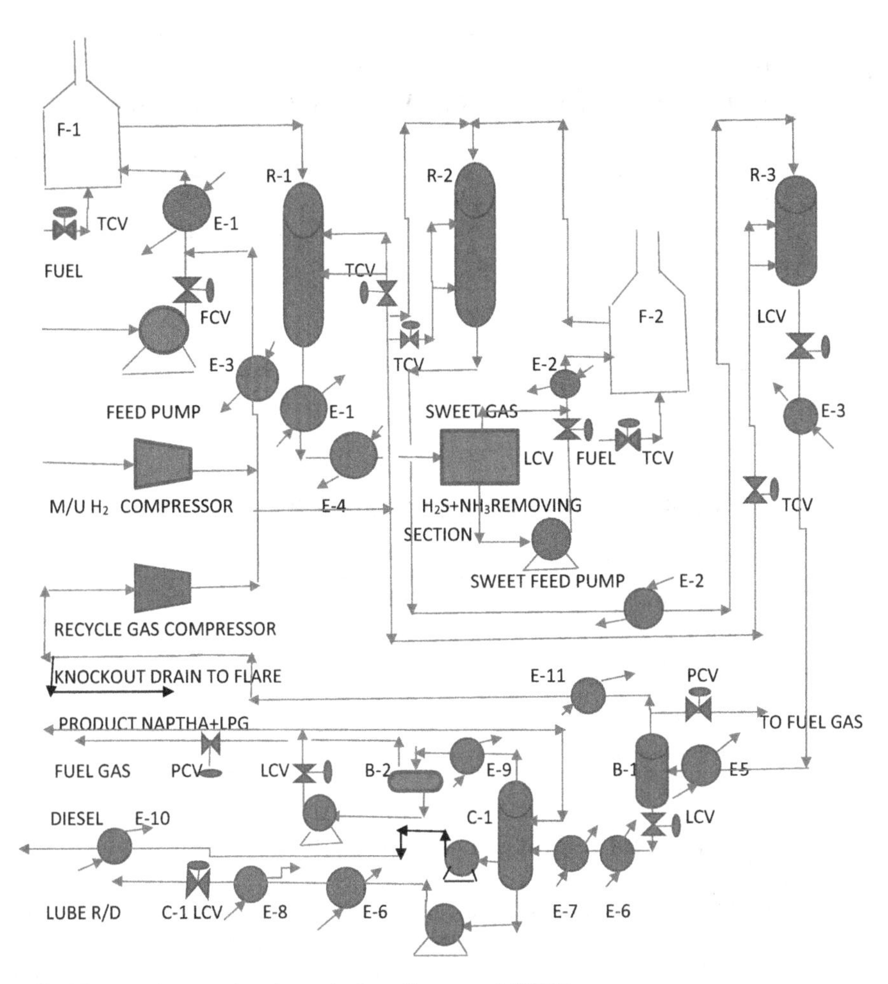

FIGURE 2.9 Case study schematic flow diagram of CIDW.

TABLE 2.18
Case Study Yield Pattern of Various Grades of LOBS ex CIDW Unit

Attribute	100N	150N	500N	150BS
H_2 consumption., %wt.	0.8	0.8	1.0	1.2
H_2S+NH_3 gen., %wt	0.8	0.8	1	2
Gas yield, %wt.	1	0.6	1	1.7
Naphtha yield, %wt.	2	1.4	3	3.5
HSD yield, %wt.	25	14	9	9
Light LOBS yield, %wt.	8	12	16	0
Main LOBS yield, %wt.	64	72	71	85

TABLE 2.19
Case Study Operating Parameters in CIDW Unit for Different Grades

Attribute	100N	150N	500N	150BS
Make Up H_2 ratio, Nm^3/m^3 of feed	50	50	50	50
Recycle Up H_2 ratio, Nm^3/m^3 of feed	150	150	150	150
HDT reactor, WABT, °C	305	310	310	315
ISDW reactor, WABT, °C	300	305	305	310
HDF reactor, WABT, °C	230	230	230	230
Reactor Pressure, bar	110	110	110	110

TABLE 2.20
Case Study Utility Consumptions in CIDW

Utilities per MT of Unit Feed Capacity	Consumption Rate
Hydrogen, Kg	7.5
LP (70 psi) steam, Kg	40
MP (200 psi) steam, Kg	50
Fuel, Kg	25
Electricity, KWH	50
Circulating Water, M^3	25

- **Metallurgy and Investment cost of CIDW unit:**

As mentioned above, the unit being a reaction unit requires three high-pressure reactors of about 110 bar pressure. The unit contains three sub-sections with two sub-sections at high pressure of about 100 to 110 bar, i.e. first one is reaction section including three reactors, two hydrogen compressors, one high pressure feed pump along with piping and exchangers, second one high pressure gas sweetening section containing amine wash column and associated high pressure pump; the third section is low pressure section of about 5 bar pressure up to a vacuum column for product drying associated with piping, heat exchangers and pumps. Due to having reaction sections at high pressure of about 110 bar and at a temperature of 300 to 340°C, alloy steel (austenitic stainless steel) in high pressure reactor, heat exchangers, furnaces are required to be used. However, the unit is a compact unit with minimum equipment unlike physical separation processes as described earlier above.

With the above, the approximate installed investment cost becomes 0.8 million USD per TMTA of unit feed capacity for 200 TMTA plant capacity.

- **Illustrations:**
  i. Q. In CIDW, VI of LOBS found to be lower than desired value though feedstock is unchanged, all operating parameters like reaction temperatures, system pressures, hydrogen flows, feed rate are normal while pour point

of the product is not given away (i.e. over specification). What should be corrective measure?

A. H_2S content of recycle gas has been checked by laboratory analysis and found to be higher at 200 ppm against normal level of 100 ppm. Hence, amine washing efficiency in recycle gas purification has been improved by increasing amine solution flow rate or improving inlet amine, i.e. lean amine strength to get back the recycle gas purity as mentioned above resulting in solving the VI issue of LOBS.
Note: like in HFU unit as discussed in sub-Section 2.4.1.4 under heading, 'Selection of catalyst', here also, minimum H_2S concentration of 50 to 100 ppm is to be maintained to avoid leaching out of sulfur in HDT catalyst as explained further in-there.

ii. Q. In CIDW, VI of LOBS found to be lower than specification, i.e. at 102 against desired value of 105 minimum while pour point is given-away, i.e. say (−)12°C against desired value of (−)9°C maximum as shown in Table 2.4. what should be remedial measure?

A. Reaction severity in isomerization reactor is reduced to increase pour point up to (−)9°C from (−)12°C when it is observed that VI of the final product also increased to a level of 107 from 102. Then reaction severity of HDT marginally reduced to avoid give-away from the desired VI of 105 to solve the overall problem.

iii. Q. How to establish the economics whether Group-I LOBS or Group-II LOBS are to be produced if aromatic extracted raffinate is used as feed?

A. Get the yield pattern in SDU from Table 2.14 and corresponding yield pattern in CIDW from Table 2.18; then get the total values of all products from market products price catalogue and calculate the values of all products plus by-product in SDU and deduct all its operating cost like solvent and utilities consumptions and costs in SDU and hydrogen and utilities consumptions and costs in HFU to get the value addition to produce Group-I LOBS. Similarly, calculate the values of all products plus by-products in CIDW and deduct the operating cost in CIDW to get the value addition in CIDW. The difference in these two value additions concludes whether Group-I or Group-II LOBS to be produced if there is freedom in market demand.

2.4.2.2 Hydrocracking and Solvent Dewaxing

We have seen in Table 2.3 of Section 2.2, i.e. in the API classification of LOBS, that there is no specification with respect to pour point. However, there are expected pour point ranges for each grade of LOBS in LOBS marketing. With this background, Group-II and Group-III LOBSs can also be produced by partial hydrocracking or hydrotreating of vacuum distillate or of raffinate produced from vacuum distillate in FEU / NMPEU, and then taking the residue, i.e. bottom of the hydrocracker / hydrotreater to SDU for physical separation of wax as discussed in the Section, 2.4.1.2 earlier.

- **Hydrocracking / Hydrotreating:**
  Principles of hydrocracking / hydrotreating have been discussed earlier in Section, 2.4.2.1. If raffinate is used as feedstock, there is no need to follow hydrocracking as LOBS would be the main product here; hence, in that case catalyst is selected accordingly to be used for reaction at lower pressure than that applied for hydrocracking, and this process is then called hydrotreating. But if vacuum distillate is used as feedstock, then generally hydrocracking, i.e. reaction at higher pressure is followed but with partial conversion, i.e. providing less residence time for the reaction. In this process, a good quantity of waxy bottom / residue would be obtained which can be subsequently processed in SDU for separation of the wax to produce finished LOBS. But due to separation of wax, instead of converting it to lube component by isomerization, the yield of the LOBS gets reduced, but still it is considered a preferred route if the refinery wants to produce quality wax which is precious to get high value in the market instead of selling low value slack wax obtained from SDU. Also, a fuel refinery which uses non-lube bearing crude oil and thus gets vacuum distillate having no lube potential thus desiring to install a hydrocracker to enhance fuel production, they need to install two stages full conversion hydrocracker to achieve the highest fuel production, but while deciding the process configuration of the refinery, if they want to make a balance of fuel and lube production, they can install a single stage hydrocracker which commercially gives rise to partial hydrocracking thereby producing a residue with lube potential because in the hydrocracking process, aromatics are converted into naphthenes; some naphthenes get cracked into waxy and non-waxy paraffins (paraffins with more than 20 carbon atoms are generally waxy), some waxy paraffins get cracked into non-waxy paraffins, and also to some extent, the normal paraffins get cracked into iso-paraffins. With reduced residence time, the boiling point of these product mixtures would remain in lube boiling range thereby preventing further cracking into fuel products.

  To summarize, if raffinate is used as feedstock, lube hydrotreater unit is to be installed and if vacuum distillate is used as feedstock, a single stage hydrocracker, i.e. partial hydrocracker is to be installed to waxy lube residue from the respective unit as selected.

  Hydrotreating, i.e. at lower pressure hydrogenation is adequate with the raffinate feedstock because the concentration of aromatics is about 25% in the raffinate as compared to about 40% in the vacuum distillate. This lower concentration of aromatics in raffinate facilitates conversion at lower pressure, i.e. at about 110 bar, thus can be processed in hydrotreater. But if vacuum distillate is used as feedstock, concentration of aromatics in the feed increases as mentioned above; hence, higher reaction pressure would be required with such feed to get the achieved conversion though it can be optimized with less residence time to get higher residue yield with lube bearing component.

  Reaction pressure should be optimized such that saturates content in the bottom residue to the hydrocracker / hydrotreater is more than 90% to finally get LOBS of Group-II or Group-III. Of course, sulfur content would always be less than 0.03% (as a requisite to be in Group-II or Group-III LOBS class) in this hydrotreating or partial hydrocracking process as its reaction is fastest among all other reactions.

- **Solvent Dewaxing of Hydrotreater / Hydrocracker bottom:**
  The hydrotreater /hydrocracker bottom is then taken to solvent dewaxing unit (SDU) to follow the chilling, crystallization of wax followed by filtration and subsequent solvent recovery and recycling as discussed in detail in the respective section earlier. In this unit, filtration rate would be higher than in the case of processing raffinate in SDU because in the raffinate there would be presence of higher concentration of aromatics (about 25%) a part of which are larger in molecular sizes (reflected by CCR properties of raffinate) thereby clogging the filter cloth pores to some extent whereas in hydrocracker /hydrotreater bottom there is very low concentration of aromatics at less than 5% found to be present.

  While concentration of wax in the raffinate is near to 25%, the concentration of wax in hydrocracker / hydrotreater bottom generally remains below 20% but it varies depending upon the selectivity of the catalyst and operating conditions of the unit.

  The dewaxed oil (DWO) obtained from this unit can be used as such as finished LOBS without further hydro finishing unlike with raffinate as feedstock of SDU because in this case the feed of SDU is already hydrotreated thereby achieving very good color and color stability. However, in the product rundown line, some anti-oxidant to be dosed in ppm range to enhance the oxidation stability as most of sulfur is removed in hydrogenation process as discussed.

  The pour point of LOBS produced would be inferior than in catalytic iso-dewaxing process as discussed in Section, 2.4.2.1 above, because in the physical separation process of SDU, there would always be cross-mixing, i.e. some wax get into oil phase and some oil get into wax phase resulting in higher pour point of DWO than in iso-waxing where the waxy paraffins which are generally straight chain paraffins get isomerized into branch chain paraffins which become non-waxy resulting in giving very low pour point of LOBS. But with higher efficiency of crystallization and filtration, and if necessary, providing recrystallization and refiltration of DWO, pour point can be further improved if not met otherwise as per Table 2.4 shown earlier.

2.4.2.3 Group-IV and V LOBS Production

- **Group-IV:**

PAO as mentioned earlier as Group-IV LOBS, and which can also be used as such as a lubricant, is a synthetic lubricant produced by chemical process with the following principles in view.

PAO is a polymer of three or four alfa decenes. Decane, we know, is a normal paraffin with ten numbers of carbon atom, and alfa decene is one unsaturation, i.e. one double bond in alfa position, i.e. at the terminal bond of carbon atom shown as follows:

$$H_2C = CH\text{-}CH_2\text{-}CH_2\text{-}CH_2\text{-}CH_2\text{-}CH_2\text{-}CH_2\text{-}CH_2\text{-}CH_3$$

While producing PAO, 3 to 4 alfa decenes get combined with the help of this double bond and thus there would no unsaturation in the polymer, i.e. PAO.

Hence, it is synthesized by chemical reaction like controlled polymerization of a pure compound. That's why it called synthetic lubricant. Specially designed reactor system has been commercialized to produce PAO.

PAO can also be semi synthetic, if the monomer decene is produced by fine cut distillation of crude oil distillate though it is rare to be present in crude oil but sometimes in the thermal or catalytic cracking process with crude oil vacuum distillates or residue, some decenes may get generated in small fractions which can be distilled out to get decene.

- **Group-V:**

All synthetic lube oils or base oils other than Group-IV lube oil (PAO) are called Group-V lube oil /base oil. Examples are: poly alkene glycol (PAG), phosphate esters, poly esters, etc. Ester means is a condensation polymer of carboxyl acid and alcohol; here, alcohol used is poly alcohol. Hence, these lube oils are also called synthetic esters. Vegetable oils also fall under this category though these can't be called synthetic lubricant and it is natural ester and basically vegetable oils are glyceride of fatty acid, i.e. it is a condensation reaction product of glycerol and fatty acid as naturally present in the vegetable oils. Generally, oils contain fatty acid of higher molecular weight which results in higher pour point, that's why most of the oils have high pour point though there are some exceptions. Also, with high content of fatty acid, organic acidity is also high in vegetable oils. Overcoming these two problems, vegetable oils become very good lubricant. As vegetable oils are produced from bio source, it is called bio-lubricant.

Synthetic esters as mentioned above are produced through condensation reaction of low concentration of carboxylic acid with poly alcohol in a specially designed unit to carry out the kinetic reaction process like synthesis of PAO as mentioned above. Low concentration of carboxylic acid helps to achieve negligible free acidity thus taking care of neutralization numbers of the lubricant while poly alcohol (also called polyol) helps to achieve higher boiling point of lubricant and lubricity as desired and the reaction product, ester has the all-lubricating properties as desired in a lubricant.

Various synthetic esters are: diesters, phthalates, dimerates, polyols, trimallitates, etc.

- **Catalyst system used for production of synthetic lubricant[6]:**

Conventionally diesters and similar other esters are manufactured by ester exchange method or by direct esterification reaction (condensation reaction) of carboxylic acid and polyol by use of catalysts like sulfuric acid, p-toluene sulfuric acid, alkyl titanate, sodium acid sulfate, ionic catalyst, etc.; this process is quite cumbersome and requires continuous monitoring to control yields and to avoid charring. The new technology has edge over this process in that no mixtures of esters are formed with new class of catalyst and a single pure liquid ester is formed with negligible free acidity. The process has less reaction time, easy to handle, energy saving with yield ranging from

95% and above. The advantage of the new catalyst is that diester can be formed both from primary and secondary alcohol. The polyol esters of tri-methylol propane and pentaerythritol with fatty acid ranging from C_5 to C_{18} carbon atoms have been synthesized.

- **Method of reaction[7]:**

One mole of acid (sebacic acid, adipic acid, azelaic acid or oxalic acid and 2.2 moles of primary or secondary alcohol (2-ethyl-hexyl-alcohol, isodecyl alcohol and capryl alcohol) and about 3% of new catalyst system were mixed and heated on electric heating mantle. The contents were refluxed for 1–2 hours depending on nature of acid and alcohol. The water formed consequent to reaction was removed by use of Dean-Stark apparatus. On complete removal of water, reaction was over and the contents were further heated for 1–2 hours more. Later on, the contents were cooled, filtered and excess amount of alcohol was removed by evaporation under reduced pressure.

2.5 APPLICATION OF LOBS

LOBSs, we know that are used in the blend to prepare the lubricant for end use in automotive engines, industrial engines, and turbines. Now, there are different sub-category of engines in each class, for example, in automotive, there are LMV, MMV, and HMV; even in these sub-categories, there are further sub classes of automotives like for specially designed automotives with unique advanced technologies and a special class of lubricant t can be used. So, there would be a lot of number of grades of lubricant.

By studying all varieties of engines, it has been found by the lubricant industry that lubricants can be broadly classified into light lube, medium light lube and heavy lube with respect to kinematic viscosity. Also, with respect to speed of the machines, friction coefficient tolerance, there are other classification of the lubricant, i.e. low VI, medium VI, and high VI lubricant. Similarly, with respect to climatic temperature conditions for the engines to be operated in, the lubricants are classified as low pour point lube, medium low pour point lube, and high pour point lubes; but now-a-days no lubricant is used for single temperature climate, rather lubricant should be of multigrade quality, i.e. same lubricant can be used in summer as well as in winter with almost same efficiency.

Lubricant is generally a blend of so many components to achieve all the properties required by the machines, but LOBS is the basic component in the blend and others are various additives added in the blend in small doses to compensate the short fall in any property that could not achieved by the LOBS only.

To produce a lubricant of particular grade, not necessarily a particular grade of LOBS, viz. 150N or 500N or 800N Or 1300N would be used, rather it may be that a mixture of two or three grades of LOBS as mentioned above are used in a fixed proportion to produce lubricant of particular grade. However, Group-IV and V LOBS are found to be very suitable to use for end use as lubricant without any blending.

Regarding functions of various additives in the lube blend, the subject is discussed in detail in Chapter 3 on Lubricant.

As far as automobiles are concerned, it can be said 150N or 150SN is used mainly for small vehicle like two-wheeler, and 500N or 500SN is used for four-wheelers whereas 150BS grade is used in locomotives, aeronautics, etc.

Similarly, would be the case for industrial engines and turbines.

The above-mentioned example of application of LOBSs, however, is not sacrosanct, but it gives a guideline for application of LOBSs.

Regarding environment point view, biodegradability is the main issue for use of LOBS, and in that sense, Group-I LOBS has least biodegradability, and it increases with increase in Group numbers with highest biodegradability in Group V LOBS; this higher biodegradability facilitates easy disposal of spent lube. However, Group-IV LOBS have more or less same level of biodegradability to that of Group-III.

Also, from coefficient of friction or VI point of view, Group-II, III and IV are more preferred than Group-I, that's why, N category of LOBS have gained more popularity than SN grade.

Another disadvantage of using SN grade LOBS in the lubricant is that due to higher presence of aromatics in the SN category LOBSs, the aromatics accelerate sludge formation in the engines thereby reduces life cycle of the engines as well as lubricant.

3 Lubricant

It is worth repeating that the lube base or lubricant base oil or base oil as it is commonly called as covered in Chapter 2 is the main constituent of lubricant. Lubricant can be broadly classified as followed:

- Lube oil
- Grease.

3.1 LUBE OIL

Lube oil is liquid lubricant whereas grease is a semi-solid lubricant. Lube oils are used in high speed engines whereas greases are used in comparatively slow speed engines or slow / intermittent moving parts of the engines.

Lube oils contain mainly base oils but with a very small addition of additive(s) as per formulation(s) developed with 'tribology' study to use the lube oil in an engine of particular design and specifications. A particular grade of lube oil can also be made with a mix of different grades of base oil instead of using a single grade of base oil based on formulation as mentioned above.

Grease on the other hand contains less quantity of base oil(s) which is/are mixed with a good proportion of one or two solid diluent like soap with or without additive(s) as per formulation as mentioned above; the solid diluent(s) as mentioned above have lower contribution as compared to base oils to the overall lubricity of the grease.

3.1.1 Lube Oil Properties and Significance

Lubricating oil has to perform many functions in an IC (internal combustion) engine such as:

- Minimize friction, i.e. wear.
- Act as cooling agent.
- Act as cleansing agent.
- Prevent rusting and corrosion of machine parts in contact.

DOI: 10.1201/9781003334422-3

- Act as sealing agent.
- Provide easy engine start and low temperature fluidity.
- Provide fuel economy.
- Eco-friendly.

Lube oil physico-chemical properties are similar to those as given in Table 2.4 and Table 2.5 in Chapter 2 as the lube oils contain mainly base oils as mentioned above. The significance of the respective properties has also been discussed in Chapter 2. Hence, it can be said that lube oil properties are influenced mainly by the base stock properties as mentioned in Chapter 2.

But to meet the requirement of the machines in the industry so as to achieve the functions of lubricating oil as mentioned above, there are other properties as explained in Webster's model to be ensured in the lube oil as follows:

- Machine seal compatibility.
- Prevention of deposit formation.
- Prevention of foaming.
- Electrical insulation properties through oxidation stability.
- Chemical / thermal stability.
- Prevention of wear.
- Emulsibility
- Prevention of energy losses.

Various lube oil marketing companies have developed lube oil(s) for each category of the machines as internationally standardized in automobile and industrial machines to ensure enhanced life cycle of the machines to make their products (lube oils) acceptable in the market. Hence, the lube oil marketing companies developed formulations, to enhance some properties as shown in Table 2.4 and 2.5 in Chapter 2 and also to incorporate some additional properties as mentioned above as required for each category of machines, using base stock(s) as key constituent and various additive compounds with objectives of the additives to enhance the respective properties of lube oil and accordingly the additives have named / titled. Hence, additives are classified as shown in Table 3.1 and Table 3.2 with respect to different objectives they contribute to lube oil.

While basic properties like viscosity, VI (viscosity index), pour point, ASTM color, CCR (carbon Conradson residue), oxidation stability, volatility for the base oils, as discussed in Chapter 2 earlier, are most significant properties to be maintained in lube oil thus making base oils as the key constituent of the lube oils, lube oils also should have some other additional properties like, dispersion, detergent, anti-wear property, anti-rust property, anti-corrosion property, antifoam property, thermal stability, improved pressure characterization property, etc. as well as enhanced basic properties of base oils as discussed above by addition of VII (viscosity index improver or friction improver), pour point improver (PPI), etc. which are achieved by addition of additives in the base oil. Regarding pour point improvement, it is always tried to achieve the pour point in the base oil manufacturing itself rather than adding

TABLE 3.1
Additive Classification for Automotive Lubricants

Surface protective additives	Performance additives	Lubricant protective additives
Anti-wear agent	Pour point Depressant	Antifoaming agent
Corrosion inhibitor	Seal oil agent	Anti-oxidant
Detergent	Viscosity index Improver (VII)	Metal Deactivator
Dispersant		
Extreme Pressure (EP)		
Friction modifiers / VII		
Rust inhibitor		

Source: [5].

TABLE 3.2
Additive Classification for Industrial Lubricants

Surface protective additives	Performance additives	Lubricant protective additives
Anti-wear agent	Pour point Depressant	Antifoaming agent
Corrosion inhibitor	Emulsifier	Anti-oxidant
Oiliness agent	De-emulsifier	Bactericides
Detergent /Dispersant	Viscosity index Improver (VII)	Bacteriostats
EP agent	Tackiness agent	fungicides
Friction modifiers / VII		
Rust inhibitor		

Source: [5].

pour point improver in the lube oil. As the basic properties and their significance have already been discussed in Chapter 2, the additional requisite properties of lube oil along with VII and PPI to be contributed by additives as mentioned above are discussed with their significance as follows.

- Dispersant:
  These are non-metallic compounds having a polar group attached to hydrocarbon chain. The hydrocarbon chain is highly soluble in base oils while the polar function attracts potentially insoluble polar contaminants. This helps in keeping the foreign bodies, i.e. contaminants in suspension with the lubricant and preventing them from agglomerating into larger particles or collecting in engine parts forming sludge and varnish. Quality or contribution of the dispersant is represented by dispersion stability and is measured by ultraviolet-visible spectro-photometer (UV-Vis) and zeta-potentiometer. In the measurement, a

relative concentration of 1.0 indicates that there is no particle deposition in the lubricant solution and dispersion stability is excellent.

- Detergent:

It keeps insoluble combustion and oxide products in suspension and dispersed. Detergents are metal containing cleanliness agents. They are the reaction products of an acidic, oil soluble hydrocarbon substrate and a metal containing base. Detergents can vary from neutral soaps to highly basic materials. Detergents are classified in terms of their alkalinity. A 300 TBN detergent added to a lube base oil at 1.67% by weight would provide the lube oil with 5 TBN of base reserve. TBN describes the neutralizing power of the detergent on weight basis.

Detergent soap is important in maintaining high temperature engine cleanliness. Basic additives (metallic base) neutralize the acidic blow by-products and acidic oil degradation products. Neutralization of acidic contaminants also helps in preventing rust and corrosion.

- All the detergent additives fall in the following categories:
- Normal and basic metal salts (oil soluble sulfonic acids).
- Normal and basic salt of high alkyl substituted benzene sulfonic acid.
- Normal and metal salt's phenates or alkyl phenol.
- Metal salt of acidic materials.
- Ashless detergents.

The performance of the detergent additive is measured by the TBN value of the lube oil. It is understood even with the base oil having very good oxidation stability, the lube oil produced from it would lead to oxidation on prolonged use before its drain cycle. This oxidation would lead to formation of acidic compound in the lube oil thereby reducing the net TBN of the lube oil. Hence, to assess the performance of the lube oil, from time to time its TBN is measured while in use in the machine. If the initial value of TBN at 5 as mentioned above doesn't deteriorate to zero or the lube oil doesn't result in a negative value of TBN before the drain cycle of the lube oil, then the lube oil so produced can be tagged to have excellent detergent property.

- Antifoamant:

Foaming in the lube oil may occur while is use in high-speed machine causing huge agitation in the lube oil. Also, the additive used in the lube oil to improve the respective property can also cause foaming, e.g. metal sulfonate. Similarly, the lube oil which is viscous can also cause foaming while in use. Foaming results in two problems, viz. misrepresents the level of the lube oil in the chamber connected with the machine, and fails to maintain proper film between two parts of the machine. Most widely used antifoam additive is silicone polymer compound from effectiveness and economics point of view. The common dosage of this additive to lube base oil is 10 to 15 ppm (parts per million) by weight depending, however, on the 'tribology' study to apply the formulation in a specific category of machine.

- Anti-wear additive / corrosion inhibitor:

Anti-wear additive protects rubbing surfaces operating with thin films, boundary lubrication. Corrosion inhibitor protects against chemical attack on alloy bearings and other metal surfaces. There may be different types of wear in the machine, viz. corrosive wear, mechanical wear, abrasive wear, fatigue wear, etc.; out of these, corrosive wear and mechanical wear can be controlled by adding additive(s) as developed in 'tribology'. Corrosive wear, i.e. rusting, is caused in the engine by the blow of byproduct(s) from the combustion chamber, mainly in the piston ring and cylinder. Corrosive wear is also caused in the diesel engine by the high sulfur content in the diesel which on combustion results in the formation of SO_2-> H_2SO_3-> H_2SO_4 thereby causing corrosion in machine parts in addition to causing pollution in the environment with the exit of flue gas from the machine. This corrosion is controlled by adding neutralizing agent in the lube oil, say, metal sulfonate; but the problem of using metal sulfonate has been discussed above in getting antifoam property of the lube oil; however, the metal sulfonate dispersed in calcium or barium hydroxide and in very small dosage can avoid a problem like foaming. Under thin film condition, more wear results in, hence, antifoaming property once achieved in the lube oil also facilitates to avoid thin film due to foam in between instead of a continuous film with right thickness. The mechanism of working of anti-wear additive is by the chemisorption of the additive on the metal surface of the machine.

As discussed above, it can be said that anti-wear additive should be polar compound with base/alkalinity in nature and example is: zinc alkyl dithiophosphate (ZDP). A detergent additive with good base/alkalinity can also act as an anti-wear additive.

- VII:

The chemistry and function of VII has been discussed earlier in Chapter 1. Here, let us highlight the available VII in the market.

VII additives are high molecular weight polymers such poly-isobutene, polyesters, poly alkyl methacrylate with molecular weight ranging from 50,000 to 1,000,000. Choice of selecting the right VII depends upon its resistance to shear and oxidative or thermal breakdown.

[Viscosity friction reduction and wear scar reduction by about 50% even at increased temperature during operation of the machine have been achieved along with oxidation stability improvement by about 6% with the use of these additives.][6]

Examples of VII widely used in the industry are: molybdenum dialkyl thio phosphate (MoDTP), molybdenum dialkyl thio carbamate (MoDTC) and similar boron derivative have been found best performing in this regard when added to a base oil. But MoDTP is found to exhibit better performance than MoDTC. These two compounds have no adverse additive-additive interference as observed.

[Compared to Molybdenum additive, boron additive reduces friction by about 25%.][6]

[In additive formulation, it has been found that at a 0.1% Mo level, in combination with boron additive with boron concentration at about 0.02 to 0.06%, the final lubricant exhibit best anti-friction property while compromising wear resistance to some extent.][6]

[The saturation in shear stability Index (SSI) is generally found at a concentration of about 3% of VM (viscosity modifier) but percent shear loss (PSL) may be not same at that level and rather found to be varying /increasing with change of the base oil from lighter to heavier grade.][7] The VM is also called viscosity index improver (VII).

- PPI:
  Lube base oils, produced from mineral oil like crude oil source as discussed in Chapter 2, contain waxy paraffin, and hence while removing these waxy paraffins in the production process there is a trade off to its removal extent in order to maintain other properties of base oils. Due to this, most of the Group-I base oils have pour point not less than (−)6°C. and Group-II and III base oils not less than (−)10 to (−)15°C. As Group-I, II and III base oils are available in abundance in the market, hence, these are the base oils used in the production of lube oils for most of the machines, and in some cases where very low pour point is required in the lube oils due to their use in very cold climate, then some PPI is to be added in the base oil to get the desired pour point of the lube oil. Examples of available PPI in the market are:
  - Chlorinated wax condensate.
  - Alkylated wax phenol.
  - Polymethyl methacrylate.
- Rust inhibitor:
  It eliminates rusting in presence of water or moisture. Rusting generally occurs during idling of the machine when there is chance of air ingress; there may be accidental moisture ingress in the lube oil from ambience or from any cooler leakage. We know that rust may occur when iron, water, and air are present simultaneously; idling of the machine may create this situation. Moreover, presence of any by-chance oxidation product which is acidic in nature can accelerate the rusting process in addition to galvanic corrosion. So, addition of rust inhibitor and corrosion inhibitor can take care of this unwarranted situation.
- De-emulsifier:
  In lube oil used in steam turbine, there may be ingress of moisture in the oil resulting in oil-water emulsion formation thereby affecting lubrication in boundary conditions. Hence, in such lube oil, a de-emulsifying additive is added to prevent this emulsion formation.

While technology cares are taken in the base oil production to ensure good oxidation stability in it as mentioned in Chapter 2, it can take care of the same on long storage of the base oil; but when lube oil produced from this base oil is used for lubrication in the machine, oxidation stability of the lube oil may not remain intact during its drain cycle; hence, additional anti-oxidant becomes necessary to be added in the lube oil in the minute proportion as guided by the 'tribology' before selling the same in the market. The various anti-oxidants available in the industry are given as follows:

- Phenolic type – as chain inhibitor.
- Aromatic amines – as free radical destroyers.

- Phosphatic sulfurized terpenes or zinc alkyl thiophosphate – as peroxide destroyers.

There are other additives are sometimes added in the lube oil, viz. thermal stability additive and extreme pressure characterization additive.

Lube oils produced out of Group-I base oils contain a considerable concentration of aromatic compounds unlike Group-II base oils. Aromatic compounds tend to polymerize in the long run when temperature of the oil rises and due to poor thermal stability of high molecular weight aromatics, these polymerize and finally decompose to form carbonaceous sludge. Similarly in extreme pressure conditions, lube oil characterization may also get changed and affect lubrication; there are additives which impart improved extreme pressure characterization to lube oils.

3.1.2 Lube Oil Classifications, Compositions and Applications

3.1.2.1 Classifications

Various grades of lube oil have been developed in the industry based on requirement of the machine; hence, lube oils are classified broadly in the names of the engines and accordingly, lube oils are classified as follows:

- Automotive lubricating oils:
  These are also of two sub categories, viz. four-stroke engine oil and two-stroke engine oil, but, in general, automotive lubricating oils are sub classified as follows:
  - Crankcase oils.
  - Universal tractor oils.
  - Agriculture pump set oils.
  - Two-stroke engine oils.
  - Gas engine oils.
  - Gear oils.
  - Brake oils.
  - Shock absorber oils.
  - Preservative cum running-in oils.
  - Automatic transmission fluids.
  - Radiator coolant.
  - Calibration fluids.
- Rail road oils:
  This lube oil has been developed to meet the diesel engine working on the principle named 'Rail Road Mechanism in auto ignition engine'. With the limited development in this category of engine, no further sub classification developed by most of the oil marketing companies barring a few.
- Industrial lubricating oils:
  Industrial lubricating oils are sub classified as follows:
  - Turbine oils.
  - Hydraulic fluids.

        - Control valve hydraulic fluids.
        - Circulation and hydraulic oils (R and O type-rust and oxidation prone).
        - Circulation and hydraulic oils (anti-wear type 1).
        - Circulation and hydraulic oils (anti-wear type 2).
        - Special purpose hydraulic oils (anti-wear type).
        - Fire resistant hydraulic oils (dilute emulsion type).
        - Fire resistant hydraulic oils (water-glycol type).
    - Spindle oils.
    - Machinery oils.
    - Textile oils (scorable type).
    - Textile oils (biodegradable and eco-friendly).
    - Gear oils.
    - Straight mineral oils.
    - Morgan bearing oils.
    - Compressor oils.
    - Refrigeration compressor Oils.
    - Semi-synthetic refrigeration compressor oils.
    - Synthetic compressor oils.
    - Stationary diesel engine oils.
    - Vacuum pump oils.
    - Gas holder oils.
    - Machine tool way oils.
    - Hydraulic cum machine tool way oils.
    - Pneumatic tool oils.
    - Steam cylinder oils.
    - Sugar mill roll bearing oils.
    - Bearing oils (compounded type).
    - Open gear compound oils.
    - Special open gear compound oils.
    - General purpose machinery oils.
    - Axle oils.
- Metal working oils:
- Metal working oils are sub classified as follows:
    - Soluble cutting oils.
    - Synthetic soluble cutting oils.
    - Neat cutting oils.
    - Neat honing Oils.
    - Spark erosion fluids.
    - Strip grinding oils.
    - Atomizing rolling oils (soluble).
    - Aluminum rolling oils (neat).
    - Steel rolling oil (soluble).
    - Steel rolling oils (neat).
    - Steel rolling compound (soluble).

- Industrial specialty oils:
- These are also sub classified as below:
  - Quenching oils.
  - Mar quenching oils.
  - Heat transfer fluids.
  - Rust prevention fluids.
  - Rubber process oils.
  - Moulding oils.
  - Agriculture spray oils.
  - Industrial white oils.
- Marine lubricating oils:
  - Engine oils (crosshead and trunk piston type).
  - Turbine oils.
  - Hydraulic oils.
  - Gear oils.
  - Steam cylinder oils.
  - Stern tube oils.

3.1.2.2 Compositions and Applications

Applications of major lubricating oils have been discussed as follows:

- Automotive lubricating oils:
  - Crankcase oils:

Automotive engines, also called internal combustion (IC) engines, are of two types, viz. spark ignition combustion engine where gasoline or LPG are used as combustion fuel and compression ignition combustion engine where diesel is used as combustion fuel. Spark ignition combustion engines are also of two sub categories, viz. two-stroke engines and four-stroke engines. Lube oils used in all these engines are called automotive lubricating oils, and also standard specifications have been developed for automotive lubricating oils for each of these three categories of automotive engines.

two-stroke engines are used in two-wheeler vehicles and four-stroke engines are used for four-wheeler vehicles. Lubricating oils in two-stroke engines are added to the fuel used for combustion; hence, in two-stroke engines, there are tremendous vehicular pollution caused by the lubricant burning along with main fuel with pollutant exiting from the vehicle exhaust. To reduce pollution from two-wheeler, development in auto engines has taken place, and now-a-days, most of the two-wheeler have into market with four-stroke engines where lubricant is not added to the fuel and instead provided in the crankcase of the engines. In the four-stroke engines, irrespective of spark ignition or compression ignition engines, crankcase is placed below the cylinder block (containing piston and accessories, combustion chamber and accessories along with intake accessories, spark-plug and exhaust accessories), and bottom of the crankcase is called lube oil pan.

The crankcase housing contains crankshaft, crank, crank pin, and connecting rod. Lubricating oil from the lube oil pan is pumped by an oil pump to provide lubrication to all parts of the crankcase as mentioned above.

TABLE 3.3
Estimate of Temperature and Shear Rate in Gasoline and Diesel Engines

Estimate of temp. and shear rate in engines	Gasoline		Diesel	
	°C	Shear rate, s^{-1}	°C	Shear rate, s^{-1}
Crankshaft Bearing	100–160	10^{-4} to 10^{-7}	110–160	10^{-4} to 10^{-7}
Piston Rings / Cylinder Bores	150–250	10^{-4} to 10^{-7}	200–360	10^{-3} to 10^{-7}
Cams/Cam Followers	60–160	10^{-5} to 10^{-8}		

Source: [8].

The automotive lubricant not only provides the desired life to the automotive parts but also contribute in reduction of fuel consumption to a large extent by ensuring proper lubrication of the said parts.

The lubricant contribution to fuel economy is reflected by the limitations of kinematic viscosity @100°C in order to predict lubricant performance. A typical estimate of temperature and shear rate in engines as confirmed in ASTM 1977 questionnaire[1] and is reproduced in Table 3.3.

[In auto engines, about 20 to 25% of useable power generated by combustion is dissipated in overcoming internal engine friction; some two-thirds of this is estimated to be due to viscous effects and the remaining one third due to boundary friction.][1] Reducing viscosity of an engine oil in the search for fuel economy may reduce hydrodynamic protection of the bearing.

The lubricant industry has testing standards for energy conserving automotive lubricant since 1983 as shown in Table 3.4.

In developing optimized fuel efficiency engine, several aspects of the engine oil must be taken into consideration as described below:

- Viscosity modifier has little effect on fuel economy as measured by sequence VIA[1].
- Base oil type has a significant impact on fuel economy, with cost-quality optimum oil being API Group-II or III base oil.
- Viscosity grade has the largest impact on fuel economy, with the lighter grades offering the most economy.
- When developing formulation within a viscosity grade following should be in consideration:
  - Friction modifiers (VII) have a beneficial effect on fuel economy on case-to-case basis.
  - Increasing the level of ZDP (zinc di-alkyl di-thiophosphate) as anti-wear additive is beneficial to fuel economy.
  - [Secondary ZDPs are most beneficial than primary ZDPs, and within the class of secondary ZDP, lower molecular weight species are more beneficial than lighter ones.][1]

TABLE 3.4
Testing Standards in Lubricant Industry

ASTM Five Car Test	1983
ASTM Sequence VI	1988
ASTM Sequence VIA	1996

Note: Subsequent development not included in the above table.

In lubricant formulation for diesel engines, a greater number of additives are found to be required than for gasoline engine. Accordingly, a fine balance and optimization of nature and quantity of various additive chemistries which go into the lubricant hold the key for sustained deliverance of high performance. As regard to use of detergent as additive in the formulation, TBN level of ten found to be optimum in the automotive car engines. However, thermal stability with respect to TBN stabilization, i.e. less or non-oxidation of the detergent (good oxidation resistance) is one of the important parameters which in turn contribute to superior dispersion characteristics of these oils.

The use of over based detergent (to get good number TBN) along with zinc containing additive also contribute to achieve good anti-wear property, i.e. good value of wear scar diameter (WSD) in WSD testing.

Typical physico-chemical properties of four-stroke crankcase oils are shown in Figure 3.1.

- Two-stroke engine oils:

  These oils are for two or three-wheelers operated by two-stroke engines which are still being operated since inception of road transport provided automobile industries. Now-a-days, however, such two-wheeler and three-wheeler engines are taken away from the road due to severe pollution created by these vehicles with tremendous increase in the number of these vehicles. The main cause of the pollution is due to direct mixing of the lubricant in the fuel resulting in exit of unburnt combustion products along with sulfur in the engine exhaust.

  The engine oils produced for two-stroke engines are made from selected base stocks blended with additives to minimize spark plug fouling, pre-ignition, to prevent rusting, minimizing deposit formation in the engines and provide protection against seizure, scuffing, and wear. There are grades which are produced to provide additional benefit of reducing fuel consumption. The oils contain diluent for easy mixing with the fuel, gasoline and the diluent also ensures proper pumpability in the fuel injection even at very cold climate. These oils

KV, cst@100⁰C	VI	Pour Point, ⁰C	Flasht Point, ⁰C
4-6.5	>90	(-)6	190

FIGURE 3.1 Physico-chemical properties of four-stroke crankcase oil.

also can be used in sprayers, small pumps, small generators, lawn movers, etc. The dosage of these oils in the fuel are very small in quantity to restrict pollution as mentioned above; the dosage, generally, remains in the range of 10 to 20 ml per liter of gasoline.

There are new generation grades of this oil produced from semi-synthetic base stocks which have been specially developed to meet lubrication requirement for high performance two engines along with very low sulfur content and negligible aromaticity in the base stocks to control air pollution from the engine exhaust; hence, Group-II or Group-III and even partly Group-IV (synthetic base stock) are used in the formulation to develop this oil. The special quality can be achieved even without mixing the synthetic base stock in the blend as mentioned above. The high performance ensures spark plug cleanliness, excellent lubricity in the engine and reduced fuel consumption. The dosage of the oils should be about 20 ml per liter of the fuel.

Typical physico-chemical properties of two-stroke engine oils are shown in Figure 3.2.

- Gear oils:
  Gear oil is also provided in a separate pan of the vehicle to lubricate its gear components to ensure life of the gears and also to contribute in the fuel economy of the vehicle. Auto industry is concerned with thermal and oxidative stability of the oil while to achieve gear surface protection and oil seal life in the gear assembly.

It is experienced that fuel contributes about 60% on fuel economy while there are other factors as shown in Figure 3.3.

From Figure 3.3, it is evident that auto industry would always prefer to reduce fuel consumption as top most priority, and in this endevour improved aerodynamics plays an important role; improved aerdynamics means reduction of ambient air flow across the moving vehicle, but this in turn increases the temperature of the gear box. Hence,

KV, cst@100⁰C	VI	Pour Point, ⁰C	Flasht Point, ⁰C
5-7.5	>95	(-)6	>75

FIGURE 3.2 Physico-chemical properties of two-stroke crankcase oil.

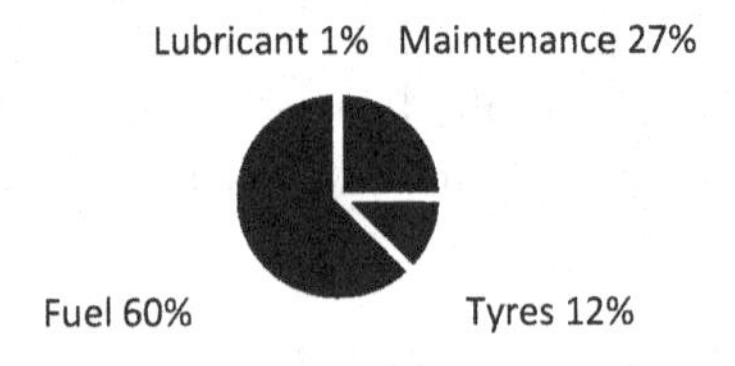

FIGURE 3.3 Contributors to fuel economy.

Source: [9].

the gear oil used should be of very good thermal and oxidative stability to overcome high temperature problem.

To meet the regulation with respect to noise reduction also, the gear cant be cooled to the desired extent; hence, enhanced thermal stability and oxidative stability of the gear oil can accommodate operating gear oil at higher temperature.

Also, to use the gear oil in cold climate, gear oil viscosity should be lower but again there becomes a problem of shear stability of the gear oil at increased temperature of the oil during motion of the vehicle; to overcome this, multigrade gear oil should be used, i.e. using low viscosity oil as the base lubricant, visosity modier, i.e. friction improver, i.e. viscosity index improver (VII) should be added in the formulation to ensure shear stability at higher temperature which also helps ensuring better wear protection.

The above qualities of the gear oil also contribute interms of longer drain interval thereby reducing maintenance cost. Typical physico-chemical properties of two-stroke engine oils are as follows.

Typical physico-chemical properties of gear oils are shown as Figure 3.4.

- Brake oils:
  It is developed from very high viscosity of about 1500 cst. @100°C (max.) but with very low pour point of about (−)40°C. In the additive package, it contains anti-oxidant, metal de-activator and corrosion inhibitor to ensure long life of brake oil and brake components. It has a high boiling point of about 200°C which doesn't give vapor lock problem in high temperature operating conditions. It is recommended for modern hydraulic braking systems including disc type.
- Gas engine oils:
  These oils are developed from medium viscosity base stocks added with ashless-detergent additive, with oxidation and corrosion inhibitor and with anti-wear additive. The oils provide excellent dispersions, high temperature detergency and oxidation stability along with corrosion protection. These oils ensure negligible carbon and ash deposits in the engine parts thereby ensuring life of the machine, and also provide fuel economy by avoiding carbon deposit (which is a result of incomplete combustion).
- Auto transmission fluids:
  There are various sub grades of these oils produced for use in specific machines / equipment as auto transmission fluids. These oils are low viscosity oils but with very high VI of about 160; hence, in the formulation, even if mineral oil base stock of Group-I category is used where VI can't be achieved above 95 to 98, right proportion VI improver to be used in the blend. Now-a-days, however, with the availability of base stocks of Group-II and Group-III category, very

KV, cst@100°C	VI	Pour Point, °C	Flasht Point, °C
13-15.5	90	(-)6	>200

FIGURE 3.4 Physico-chemical properties of gear oil.

negligible amount of VI improver is required to be added in the blend. In the formulation, additives like anti-oxidant, detergent-dispersant, anti-wear, anti-corrosion, and antifoam are also used.

Different grades of low viscosities of about 5 to 10 cst. @100°C and high VI up to 120 are produced in this group to use in automatic transmission and power steering units in automobile and light trucks, for synchro-mesh gears, and in all types of rubber seal. There are grades produced which can be used in all commercial power shaft transmission, industrial torque converter and in automatic transmission. These oils also require the above-mentioned additives in the formulation, but the formulation also require protection against copper corrosion.

Some grades are produced for hydraulic systems fitted in tractors and other farm equipment.

There are grades also produced for use in turbo transmission systems and industrial torque converter where in the blend EP additive is also used in addition to above mentioned additives; the oils can also be used in fluid couplings, hydraulic systems and gears, and thus get wide use in the railways.

- Radiator coolant:
  It has a freezing point of about (–)34°C min. it is generally made from synthetic base oils rather than from mineral base oils. Its formulation includes corrosion inhibitor, antifoam, and a dye to demarcate the product. It has high boiling point along with its low pour point as mentioned above which prevent its boiling in summer as well as freezing in winter or in a cold location. It is recommended for use in passenger cars, SUV, three-wheelers, diesel engines, etc. it protects non-metallic components like gaskets, hoses and thrust seals in the cooling circuit. It need not be used 100% as such, rather it can be diluted with distilled water; to get best performance, an optimum ratio of 30:70 between coolant and water has been recommended by many companies.
- Shock absorber oils:
  These oils are made up from base stocks of low viscosity like 10 to 18 cst @40°C, and with low pour point of about (–)30 to (–)45°C; sometimes higher viscosity of about 40 cst also used to make special grade of shock absorber oil. Oils produced from these base stocks containing zinc type anti-wear additives to reduce wear. These oils have excellent damping characteristics. Some grades also use phosphorous containing anti-wear additives for better wear protection. These oils also have high VI, oxidation stability and antifoam characteristics by use of desired additives. Some grades have excellent seal compatibility and damping effect to increase the life of shock absorbers. Some grades with low sulfur content provide superior performance.
- Rail road oils:
  This is an extra performance generation-IV lube oil used in diesel engines and with a specific grade. It has high viscosity index (VI), excellent detergency-dispersions, good filterability and excellent oxidation stability. It has high level of alkalinity reserve, i.e. with high TBN, to protect engine parts against ill effects of sulfur in the fuel over long service period. It is recommended for ALCO locomotives also, but it not recommended for diesel engine fitted with silver lined bearings such as G.M. locomotives. To use in G.M. locomotive,

special grade to be made to take care of silver lined bearings, for example, in that case, very low level of zinc in the additive package is to be maintained.

- Industrial lubricating oils:
  Industrial lubricants are used in various category of machines as mentioned in sub-Section 3.1.2.1. To make the lubricants, selectively refined base stocks containing anti-oxidant, rust-inhibitor, and antifoam as additives are used to achieve best possible lubrication performance. Applications are discussed for different sub-categories as follows.
  - Turbine oils:
    They exhibit oxidation stability of highest order, excellent de-emulsibility and resistance to foaming along with ability to release entrained air quickly. Oils specifically made can be used for steam turbine, gas turbines and hydraulic turbines. With addition of anti-wear additives, the oil can be used for geared turbines. The oils with added quality of lower pour point of about (–)30°C can be used as refrigeration compressor oil.
    Now-a-days, ashless industrial oil formulations have come into place. Here, in the formulation, the use of the additive, ZDP (zinc dialkyl di-thio phosphate) is to be avoided. This results in achieving ashless oil and in turn provides good filterability; thus, the oil can be used in hydraulic turbines. [But to replace ZDP, EP/AW (extreme pressure/anti-wear) additives are used; however, when ZDP is used in the additive package, effort should be to restrict its concentration within a maximum limit of ten ppm.][1]
    Commonly used ashless additives packages are as shown in Table 3.5.

The additives in industrial lubricants have a polar character, and they compete among themselves for the metal surface of the machine parts in contact in additive-additive interaction process.

The lubricant should retain its properties throughout its drain cycle. Lubricant should be compatible with contaminants, such as water, calcium detergent, ZDP, etc.

In steam turbine, in addition to engine lubricant, a specific lube oil is required as its governor control fluid. Fire resistant phosphate ester fluids are found suitable for

TABLE 3.5
Ashless Additive Package vs. Functions

Sl. No.	Additive(s)	Function
1	Amine/Hindered Phenol	Anti-oxidant
2	Phosphate/Phosphoro-thionates / Thio-Phosphate	Extreme Pressure / Anti-wear (EP/AW)
3	Imidazole/Alkyl Succinic acid half ester/N-Oleyl Sarcosine / Iso-Nonyl Phenoxy acetic acid	Anti-rust
4	Tri-Azole	Metal Deactivator
5	Acrylic Ester / Silicone	Antifoam
6	Poly-Glycol/Imidazoline derivative	De-emulsifier
7	Carboxylic derivative	Ashless dispersant

KV, cst@40°C	VI	Pour Point, °C	Flasht Point, °C
28.5-35	>90	(-)6	>150

FIGURE 3.5 Physico-chemical properties of turbine oil.

the purpose as compared to petroleum lubricant oil. But contamination with water in phosphate ester may cause corrosion of machine parts; hence, if such fluid is used, serious care should be taken in this regard.

Typical physico-chemical properties of Turbine oils are shown as Figure 3.5.

- Hydraulic fluids:
  Hydraulic fluids are of various sub-categories as discussed in the classification of lube oils above. Their applications are discussed as below.
  - Control valve hydraulic fluid: it is prepared from base oils having excellent lubrication characteristics with focus on anti-oxidant selected in base oil product storage and/or in final lubricant formulation. The hydraulic fluid selected should consider its application in the circuit having fine sintered bronze filters.
  - Circulating hydraulic oil (R and O type): it is rust and oxidation inhibited oil having good de-emulsibility characteristics and long service life. It is recommended for circulation system of industrial gear boxes, air compressors where turbine quality oil is required, and for plain and anti-friction bearings of turbo feed pumps, turbo blowers, etc. it is also recommended for intermittent operating turbo generator sets.
  - Circulating hydraulic oil (anti-wear type 1): here additive package on the base oil should carefully include anti-oxidant, anti-wear, anti-rust, and antifoam additives. This oil is also used for compressor crankcase lubrication but mainly recommended for lubrication of turbines and equipment having silver coated components. This oil has very long drain cycle.
  - Circulating hydraulic oil (anti-wear type 2): here, in the formulation, special care is taken to use sludge dispersant in addition to above mentioned. These oils are recommended for use in paper mills, plastic film calendars, coal pulverizers, etc. this grade is also used in compressors, machine tools, hydraulic and circulation systems and enclosed gear boxes which don't require extreme pressure (EP) lubricant.
  - Special purpose hydraulic oils (anti-wear type): these are premium hydraulic oils having excellent lubrication characteristics with anti-wear properties, excellent thermal/oxidation stability and high FZG rating, i.e. scuffing rating for industrial gear oil and automotive gear oil. These oils have good filterability and very useful for use in screw compressors. These oils have very low pour point of about, say, (−21)°C and are recommended for high performance electro-hydraulic or numerically controlled systems. These oils provide additional characteristics as follows:
    - Good film strength.
    - High viscosity index (VI).

  - Resistance to foaming.
  - Excellent corrosion resistance.
  - High de-emulsibility.

    These oils can also have low pour point like (–)39°C up to (–)24°C for application in very low operating temperature.
- Fire resistant hydraulic oils (dilute emulsion type): these oils are oil-in-water type fire resistant hydraulic fluids specially developed for hydraulically operated mining pit-props. This oil is a stable emulsion in water and provide long satisfactory service in hydraulic systems in mining sector.
- Fire resistant hydraulic fluids (water glycol type): this oil contains about 60% Glycol and 40 and water. This is fire resistant oil and contains anti-wear additive and combination of powerful anti-corrosion additives and oxidation inhibitors. This fluid is also suitable for coal mines where fire resistance is of great importance and it is also recommended for foaming services such as continuous casting machinery, etc.

Typical physico-chemical properties of hydraulic oils are shown as Figure 3.6.

- Spindle oils:

  This oil is made from low viscosity base stocks having inherent oxidation and thermal stability. These base stocks are further added with anti-oxidant, anti-wear, anti-rust, and antifoam additives to produce spindle oils. Spindle oils are recommended for bearing lubrication of high-speed textile spindles and tracer mechanism and hydraulic systems of precision machine tools. Depending upon the machinery condition and their operation, power saving of about 5 to 7% can be achieved using this oil.
- Machinery oils:

  These medium to high viscosity lubricants and contain film strength and anti-rust additives and provide good lubrication for general machinery even under boundary lubrication conditions. These oils can maintain a thin film both under light and medium loads and provide protection against corrosion and rust. These oils are recommended for all types of industrial machinery using once-through lubrication systems. These oils are also recommended for machine tools, textile machinery, and other machine parts which are to be lubricated with thin films. Heavier grades can be used for lubrication of small open gears under light duty conditions where oil is applied intermittently.
- Textile oils (scourable type):

  These oils are medium viscosity oils having good scourability. These oils minimize wear under boundary lubrication conditions, provide protection against

KV, cst@40^{0}C	VI	Pour Point, ^{0}C	Flasht Point, ^{0}C
9-11	>70	<9	>120

FIGURE 3.6 Physico-chemical properties of hydraulic oil.

rust and corrosion and minimum tendency to drip or form mist. These oils are recommended for use in looms and other textile machinery where scourable type oils are required to remove the stains from the cotton. These oils are made from light colored mineral base stocks having excellent oxidation stability with addition of additives to ensure excellent scourable properties, anti-rust, antifoam, and anti-wear properties. With specially prepared textile oils, scourability of about 99% can be achieved as compared normal value of about 27%.

- Textile oils (biodegradable):
  These oils are prepared from vegetable feedstock as base stocks instead of using mineral base stocks. Similar additive packages are to be used in these oils also to get anti-oxidant, anti-rust, antifoam, and anti-wear properties. The oils produced are found be over 95% biodegradable.
- Gear oils:
  These extreme pressure type gear oils which contain sulfur, phosphorus compounds, and have better thermal stability and oxidation stability as compared to conventional lead-naphthenate gear oils. These oils have good de-emulsibility, foam resistance, and with anti-rust and anti-corrosion properties. These oils are recommended for use in heavy duty enclosed gear drives with circulation or splash lubrication system operating under heavy or shock load conditions up to temperature of 100°C, however, for worm gears, bulk oil temperature is restricted to 95°C. These oils can be used for plain and roller bearings, sliding surfaces, chain drives, sprockets, flexible couplings employing splash or circulation lubrication system requiring EP type oils. For heavily loaded low speed open gears, gear oils with higher aromaticity (blackish) can be used but with additives like EP and anti-wear types in the formulation. Some gear oils are made with tailor made additive packages to get higher energy efficiency to ensure reduced friction, extension of gear life and prevention of irregular wear. These oils are recommended for all types enclosed gear drives, heavily loaded bearings and industrial gears running at high speeds, high pressure and under high impact load.

Typical physico-chemical properties of industrial gear oil are shown as Figure 3.7.

- Morgan bearing oils:
  These oils are made from high viscosity and high VI base oils and used as bearing oils. These oils possess excellent film strength with poor miscibility in water thus avoiding wear in roll neck bearings of still mills. As these oils meet the specifications set by Morgan Construction Company, USA, it is called Morgan bearing oils. These oils also have additional properties, viz. extreme pressure, anti-rust, anti-wear, anti-oxidant, low pour point with addition of PPD (pour point depressant), and can be used as circulating oil in the bearings.

KV, cst@40°C	VI	Pour Point, °C	Flasht Point, °C
28.5-35	90	(-)9 - (-)12	>200

FIGURE 3.7 Physico-chemical properties of industrial gear oil.

These oils have high load bearing capacity with excellent de-emulsibility characteristics. These oils are recommended for use in no-twist-rod mills and also in applications where service load conditions are encountered.

- Compressor oils:
  These oils are meant for providing satisfactory lubrication under prolonged high temperature and high load conditions. These oils are made from medium to high viscosity base oils with high VI around 95 and have thermal stability with excellent anti-oxidant and anti-rust properties by addition of additives. Also, to reduce oil consumption, special polymeric additive packages have been developed to produce special quality compressor oils. These oils are excellent cylinder lubricant for high performance reciprocating and rotary air compressors and can be used up to 220°C compressor discharge temperature. There are other kinds of compressor oils specially blended with turbine oils base stocks with careful selection of anti-oxidant, anti-rust, and antifoam additives which don't react with ammonia gas. These oils have also good de-emulsibility property and low deposit and sludge formation tendency. These oils are recommended for ammonia and syn-gas compressors in both centrifugal and screw type. There is other kinds of compressor oil prepared from selective base stocks after blending it with special compounding agent. These oils give excellent resistance to water. The carbonaceous deposits formed, if any, are soft and fluffy and have minimum tendency to adhere to metal surfaces. These oils provides excellent anti-rust and anti-corrosion properties and have good fluidity at low temperature. These kinds of oils are recommended for cylinder lubrication of single and multi-stage air compressors handling moist air. These oils form stable emulsion with water and don't affect lubrication of cylinder liners, piston rings, valves, etc.; but these kinds of compressor oils are not recommended for crankcase lubrication. These grade of oils developed from base stocks of very high viscosity even with moderate VI of 85 to 90 can be used for lubrication of syn-gas compressors if the sulfur content is low, say, 0.25% max. These oils minimizes wear of cylinder liners and rings and ensure long life of the machine.
- Refrigeration compressor oils:
  These oils are made up from base stocks with low to moderate viscosity but with very low pour point to ensure fluidity at cryogenic temperature of the fluid handled by the machine. By proper additive packages used, these oils can resist deposit formation and foaming, and exhibit good oxidation and thermal stability and ensure high condenser efficiency and reduced compressor valves maintenance due to less deposits. These oils are recommended for wide range of refrigeration compressors using all conventional refrigerants except sulfur dioxide, and are suitable for both reciprocating and rotary compressors. Specially prepared refrigeration compressor oils are recommended for lubrication of hermetically sealed refrigeration compressors. To use these oils for compressors where wear of the machines are vulnerable, anti-wear additives are used in this oil.

Typical physico-chemical properties of the oil are shown as Figure 3.8.

- Stationary diesel engines:
  These oils are non-detergent stationery engine oils developed from base stocks of moderate viscosity with good VI of about 90 added with anti-oxidant, anti-wear, and anti-corrosion additives. These oils are recommended for low and medium speed stationery diesel engines where operating conditions don't require detergent oil. These oils are found to give satisfactory performance in lubrication of reciprocating air compressors.
- Vacuum pump oils:
  These oils are developed with base stocks of moderate viscosity and VI but with low pour point of about (−15)°C but the base stock and additives should be such selected that the produced oil has very low vapor pressure, say, of the order of 1×10^{-6}mmHg at 20°C, and have good thermal stability. These oils are used for lubrication of vacuum producing pump.
- Machine tool-way oils:
  These oils are made up from medium to high viscosity base stocks but even with lower VI. These oils have mild EP, tackiness, anti-rust, and antifoam additives enabling higher loading of machine tools, minimum stick slip and chattering, and provide improved surface finish to the job. These oils protect parts against rust and corrosion, and reduce oil consumption due to good adhesive characteristics. These oils are excellent for slide way lubrication of planners, grinders, horizontal boring machines, shapers, jig borers, etc. having high accuracy work, and also recommended for lubrication of gears by oil can.
- Pneumatic tool oils:
  These oils are made up from medium viscosity base stocks even with lower VI but should have tackiness, anti-wear, and EP additives, and blended with a special compound which forms a tenacious oil film by mixing with condensed moisture. These oils have excellent load carrying ability and rust preventive properties. These oils maintain oil protective films both at low and high pressure and don't give obnoxious odor even if the equipment is operated under confined environment. These oils are used for lubrication of pneumatic equipment like rock drill, jack hammers, chippers, wagon drills, etc. and also used in small in-plant pneumatic tools like grinders, filling tools, drills, tappers, impact wrenches, etc.
- Steam cylinder oils:
  These oils are produced from high viscosity base stocks with moderate VI of about 85 to 90 to get excellent oiliness and film strength properties. These oils resist water washout and minimize deposit formation even at elevated temperature. These oils are recommended for steam engines operating under wet saturated steam conditions. These oils are widely used as calendar bearings and sugar mill roll bearings, and also can be used for worm gears. These oils are not suitable for the machines where steam condensate is re-used. Steam

KV, cst@40⁰C	VI	Pour Point, ⁰C	Flasht Point, ⁰C
28.5-35	90	(-)21	>200

FIGURE 3.8 Physico-chemical properties of refrigeration compressor oil.

cylinder oil grades produced from base stocks of very high viscosity having excellent chemical and thermal stability are recommended to reduce steam consumption due to good sealing characteristics and to minimize frictional losses. These oils are recommended as cylinder lubricant for steam engines handling superheated steam, where the condensate is separated from oil and re-used for process. These oils are not used for wet cylinder conditions. These oils are also used in oil tempering and wire drawing applications with good result.

- Bearing oils (compounded type):
  These oils are made up from base stocks having moderate high viscosities and specially developed for the lubrication of exposed parts of the machine subjected to excessive water wash in marine and industrial applications. These oils provide excellent lubricity and minimize wear under shock load conditions. These oils are used in plain bearings, gears, and slides of steam engines, punch presses, etc. where conventional lubricant can't be used. These oils also find application in certain wire drawing machines. Special compounding agent is added to these oils to produce grade to be suitable in the machine to improve load bearing characteristics and resistance to leakage of oil.
- Open gear compounds:
  These oils are produced from base stocks having very high viscosities even with high aromatic content thus becoming blackish in color. Additive like special tackiness agent which reduces oil consumption is used to produce these oils. These oils have good load carrying ability and are recommended for open gear, wire rope, and chain drive lubrication. These oils can also be used in kiln tyres operating at high temperature in cement mills and chemical plants, but additive to enhance thermal stability to be added to avoid deposits abnormally. These oils are not recommended for use in food processing industries. However, any lubricant to be used in the machines meant for food proccssing needs special clearance from the respective process licensors.
- Axle oils:
  These oils are produced from moderate viscosity base stocks even with low VI of about 40 only, but the oils should have good oxidation stability and anti-wear properties by use of proper additive package. These oils ensure long service life of the machine due to having good oxidation stability. These oils are used by railways and steel plants for lubrication of plain bearings, axle boxes of locomotives, wagon and other rolling stocks due to very good load bearing and anti-wear properties.
- Metal working oils:
  These are also called cutting fluids. Consumption of cutting fluid comes second to automatic lubricants and comes before general industrial lubricants. Cutting fluids used in machinery tools fabrication poses a challenge in the economics due to their high consumption approximate @2 liters per minute and it is also found that in the economics of making tools, its consumption contributes about 5 to 15%; hence, effort should be there to reduce its consumption near to minimum quantity lubrication (MQL) from its previous established required quantity lubrication (RQL).

Since inception, cutting fluids developed were water-miscible as water can take way most of the heat generated during tools cutting operation though the metal chips generated during cutting also take away much of the heat generated, but problems with water-miscible cutting fluids are possibility of corrosion on the machine parts used for cutting operation enhanced particularly by possible leakage in hydraulic system, gear oil system, etc. Also, due to water present in cutting fluids, it leads to water pollution caused by inadequate separation of oil from drain water in the discharge waste. With these problems, water addition to cutting fluids tried to minimize but this effort was found to create heat dissipation problem as discussed above thereby affecting finish of the tools produced as well as affecting the surfaces of the machines used. Hence, efforts continued to use low viscosity oils while reducing or removing water from the cutting fluids to be developed because low viscosity of the oil enhances the heat dissipation. But it is understood that using mineral base stocks as cutting fluids, lower viscosity means lower flash point thereby posing safety as well as high evaporation loss in using low viscosity mineral base stocks. Hence, alternative, bio-lubricant has been explored and it is found that some esters while having low viscosity have adequate high flash point, and also it prevents the problem of water pollution as discussed above. A typical comparison of low viscosity ester and conventional mineral oil-based cutting fluid is shown in Table 3.6.

Hence, now-a-days, low viscosity esters are used by many lubricant marketing companies as cutting fluids while water-miscible cutting fluids with addition of anti-rust and anti-corrosion additives in the blending package are also continued as follows.

- Soluble cutting oils:
  These are milky emulsion with water containing anti-rust and corrosion inhibitor along with a biocide to prevent bacterial growth in the emulsion. These oils are used with about 5 to 10% oil in the emulsion with oil viscosity of about 20 cst @40°C with a flash point above 150 COC °C; for grinding purposes, more dilute emulsions below 5% are recommended. For stable emulsion, oil should be added in water, and not vice-versa. The cutting fluids can be used both for ferrous and non-ferrous metal.

TABLE 3.6
Properties Comparison between Esters and Mineral Oils

Sl. No.	Bio Ester	Mineral Oil
Viscosity mm²/s, @40°C	7	7
Flash Point, °C	185	135
Evaporation loss, %wt.	20	80

Source: [10].

The above emulsions provide excellent finish, excellent cooling, and reduce the pollution of water due to addition of biocides in the additive packages.

- Synthetic soluble cutting oils:
  These are water soluble but not produced from any mineral oil or fatty matter. These oils are transparent added with anti-rust and anti-corrosion additives, which can protect cuprous metal components of machine tools, and are specially recommended for grinding operation of iron, steel, non-alloyed steel, and nickel chrome steel.
- Neat cutting oils:
  These non-staining type cutting fluids blended from high VI base stocks with low viscosity of about 10 to 40 cst. @ 40°C with flash point varying from 135 to 160°C, COC. These oils contain chlorinated fatty material enhancing its fluidity which reduces tool wear, and heat dissipation characteristics which improves machinability. These fluids are used in multi tools set ups where a variety of operations are performed for ferrous and non-ferrous tool cutting. These are used for thread grinding, form grinding, and milling using multiple cutters. These oils ensure good finishes and protect tools and machine from excessive wear. Sometimes presence of active sulfur in the cutting fluids help extra cutting assistance under continuous long work cycles. These fluids are also used in deep hole boring, deep hole drilling including gun drilling, and trepanning operations where pressurized coolant system is employed for easy swarf removal and effective cooling at the cutting point. It provides satisfactory performance in deep hole boring of stainless steel.

  Chlorinated and sulfurized fat in the cutting fluids provide anti-weld, oiliness and EP properties to the cutting fluids with contribution of chlorine to get EP properties and sulfur to get extra cutting assistance. These oils are recommended for machining operations high tensile stress stainless steel as well as nickel-chromium alloys, gear cutting, hobbing, drilling/reaming, and thread cutting. The fluids have excellent foam release ability and prevent overheating of the work-piece thus eliminating chance of metal distortion thereby ensuring good finish.
- Neat honing oils / spark erosion fluids:
  These oils are very low viscosity oils with viscosity of about 1.5 to 7 cst at 40°C, and naturally having low flash point thus resulting in high evaporation loss and taking special care on safety, these oils are used for specific purposes where other cutting fluids as mentioned above can't be used. These oils also have similar other good qualities as mentioned above with other cutting fluids.
- Strip grinding oils:
  These produced from highly refined base stocks and contain special additives. These oils don't create stain or corrosion on the machine or tool. These oils are used in some strip grinding machines.
- Aluminum rolling oils:
  These oils can be used in diluted or neat type. While soluble types are of moderate viscosity thus with higher flash point, the neat type is generally of very low viscosity of about 1.5 to 3.0 cst @40°C thus with low flash point thereby limiting the uses in respective machines.

These oils are used for the hot rolling of aluminum Properly and Morgan Rod Mills for high speed continuous operations. Periodic addition of biocide additive in the cutting fluids should be compatible with the emulsion in the cutting fluids. These oils contain anti-rust and anti-corrosion additives also.

- Steel rolling oils:
  These oils are also used in soluble form as well as in neat form. While neat forms are of moderate to lower viscosities, soluble forms can be of moderately higher viscosities. These are emulsion in water category, but due to stable emulsion, their consumption is lower than other rolling oils. Due to its proper composition with natural fats, emulsion shows excellent 'Plate out' characteristics to ensure roll and strip cooling. Proper coefficient of friction gives correct roll bite, less strip breakage, and longer roll life. The neat oils are produced from highly refined base stocks having low sulfur and having anti-wear, anti-oxidant, and lubricity improver in the additive package; while it used in all types mills including multi-roller mills, it can be additionally used for low viscosity, low pressure hydraulic systems of rolling mills to avoid contamination of cutting fluid with hydraulic oil.
- Industrial specialty oils:
  These oils are meant for use in stationary equipment, parts unlike rotary machines, and also produced for use in agriculture to prevent attack from insects, in pharmaceutical, cosmetic sectors, and in specific manufacturing process fluids to achieve the desired purposes. Various grades are discussed as follows:
- Quenching oils:
  These are low viscosity oils blended from high viscosity base stocks with VI about 90. The oils are used for quenching operations on a wide variety of steel plants such as nuts and bolts, ball bearings, brake drums, etc.

  The oils have good oxidation stability, good fluidity due to low viscosity, and low volatility with higher flash point of about 180°C (COC). The polar additive used in the oil blend contributes to its accelerated quenching property. The oils are suitable for hardening of a wide variety of high speed steel tools, ball bearings and other components.

  There are special grades prepared in this category with base stocks used of low sulfur content but with high oxidation, high thermal stability, and low volatility. In addition to above use, these oils are also used for hardening of set screws, crankshaft, axles, camshaft, steering arms, etc. These oils don't contain any fatty material, and the oils can be used for quenching components after nitriding operation in the steel plants.
- Mar-quenching oils:
  These oils have exceptional oxidation stability and thermal stability to ensure consistent quenching characteristics and longer oil life. The oils are highly suitable for quenching finished components at a temperature of about 200°C. This type of quenching provides structural especially suitable for manufacturing process like gears, tools, dies, and bearings.

- Heat transfer oils:
  These oils possess excellent oxidation and thermal stabilities, good fluidity due to low viscosity, low volatility, low vapor pressure, and good VI of about 90. The oils give long and trouble-free service life in industrial heat transfer systems. These oils are recommended for heat transfer systems up to temperature of about 250 to 300°C depending upon formulation. The grades with low sulfur content and low CCR (carbon conradson residue) values provide superior performance in quenching, and can be used in the heat transfer systems up to temperature of about 315°C.
- Rust preventive oils:
  These are oils which on application on the metal surfaces form a thin film thereby preventing air/water from coming in contact with the metal surfaces which in turn prevents the rust formation on the metal fluid; even if there is presence of moisture on the metal surfaces, air would be prevented to enter inside the oil film to form galvanic cell which causes corrosion / rusting. In preparing such oils, basically oil soluble surface-active agent is used which is dissolved in a solvent. On application on the metal surface, the solvent gets evaporated out and the film of oil remains on the metal surface where the surface-active agent passivates the metal surface by ensuring a continuous non-porous oil film all throughout the metal surface even by replacing the moisture particle, if any, from the metal surface. These oils are also called de-watering fluids. The oils can be applied at room temperature by sipping or spraying. The film formed on the metal surface can be easily removed again by a petroleum solvent. These oils protect metal surfaces even from a fingerprint.

  These oils are used for short term storage before wrapping, during idling like for the period of in-between operations. While the oil films are compatible with lubricating oil, it can be easily wiped out with the help of petroleum solvent. these oils are used by the engineering industries for protection of their ferrous components from rusting during indoor and outdoor storage.
- Rubber process oils:
  There are three types of rubber process oils, viz. naphthenic type, paraffinic type and aromatic type.

  The naphthenic type is blended from specially selected base stock (with high naphthene concentration than paraffin and least aromatic concentration) added with additives. These oils are used in the processing of natural rubber feedstock in the manufacture of automobile tyres, tubes, and many molded rubber goods. Grade prepared from these oils can also be used for dust stop lubrication in Banbury machines. These oils can be used for processing of butyl rubber.

  The paraffinic type oil is made up from base stock rich in paraffin content with low sulfur content, i.e. from Group-II base stock, added with additives. These oils also used for processing of both natural and synthetic rubber feedstock in the manufacture of automotive tyres, tubes, and special grades of rubber goods.

  The aromatic type is a black colored liquid, generally obtained as intermediate in base stock manufacturing processes, and are used for processing of rubber feedstock in the manufacture of black colored rubber goods like battery casings, rubber bushes, etc.

- Molding oils:
  These oils are very low viscosity oils produced from selected base stocks added with additives. These oils provide excellent de-molding performance with good surface finish to the molded products. The base stock so selected is non-toxic and does not form fumes during its application through spraying due to reduction of its volatility by suitable additive. These oils can be applied to aerosol systems, in the production blend of concrete sleepers. The oils can also be applied in spun pipes and other molded products.
- Agriculture spray oils:
  These oils are prepared from high quality base stock specially for protection of apple trees from a particular species of insects. The oils are sprayed in the form of oil-in-water emulsion on the trees when ambient temperature is as low as 3 to 4°C. The emulsion dissolves the waxy protective shield of the insects creating an oil film envelope around the insects thereby killing the insects cutting off air supply. The pour point of the oils should be as low as (–)6°C to ensure its application in cold climate. Similarly, these oils can also be applied for protection of eucalyptus, cinchona, etc. these oils are also non-toxic and approved for use in food research application due to non-presence of aromatic (carcinogenic) compound thereby getting USA FDA approval.

  While the above spray oils are used mainly for agricultural purposes, there are grade to use as rubber spray oil. Like agriculture spray oil, it is also very low viscosity oil added with additives. This oil is used in rubber plantations. It has excellent solvent power with copper oxy-chloride for spray on rubber plantations to prevent attack of fungus on the plantations which leads to abnormal leaves fall affecting the growth of the trees resulting in loss of latex (rubber) yield. This oil is also applied by spraying in mini micron in presence of air on the plantations. The spraying enables the copper particles to readily and uniformly distribute on the leaf surfaces and leaf stalks, and at the same time not permitting copper to be easily washed out.

  Similarly, there are grades like mango spray oils which again falls under agriculture spray oils. This oil is of a bit higher viscosity about 15 to 20 cst @40°C, and controls the deadly sooty mold diseases by killing virus magnifeare organisms. Generally, 3 to 4% oil-in-water emulsion is sprayed on the plantations. It readily emulsifies with water and forms emulsion and remains uniform over a wide range of temperature even at sub-zero temperature.

  All the agricultural spray oils are either applied with FDA approval or approval from respective competent authority of the respective countries.
- Industrial white oils:
  These oils are made up from very low viscosity base stock of Group-II, Group-III or Group-IV category, but economically available, Group-II base stocks found to be useful to make this grade of oil. The oil generally passes through FDA test due to no aromaticity in the oil. The oil blended with anti-oxidant additive can be applied in the manufacturing processes of pharmaceutical industries. The color of the oil is transparent like water, have very low pour point. The oils are also used in the manufacturing process of cosmetic industries.

- Marine lubricating oils.
  These are similar to automotive oils discussed earlier with exception that due to many of the marine engines being of old design still operating due to economies posing challenges to the lubricant manufacturers to provide lubricant to continue to get fuel economy in the older engines, particularly due to heavy load engine operation in marine where temperature of the engines remains high for longer duration unlike road transport automobiles.

As classified above, the different sub-categories of marine lubricants are discussed as follows.

- Engine oils (crosshead and trunk piston type):
  These oils are of low viscosity of about 10 to 14 cst. @100°C, moderate high VI of about 90 and with higher flash point of about 200 to 220°C and are produced by blending of highly refined base stocks with additive formulation to have excellent thermal stability, excellent water separation characteristics, and antifoam characteristics. These oils are recommended for engines having separate cylinder lubrication and hydraulic system with circulation facility.
  Grades are also produced for crankcase lubrication of trunk piston type marine diesel engines having separate cylinder and hydraulic systems as mentioned above.
  Grades are available for engines operating with fuels up to 1% sulfur content.
  Oils with higher TBN value are also available for the engines operating with fuel up to 2% sulfur content and sometimes above.
- Marine turbine oils:
  These oils are blended with specially selected anti-oxidant, anti-rust, and antifoam additives. These oils are recommended for all makes of turbines, reduction gears, auxiliary turbine installations and hydraulic system with circulation facility.
- Hydraulic oils:
  Oils specially meant for this purpose are made similarly like above except with different formulation but with same additives, viz. anti-oxidant, anti-wear, anticorrosion, and antifoam. Different grades are produced in this category based on naphthenic base stocks and paraffinic base stocks with higher VI. These oils are recommended for hydraulic systems of deck machinery, steering gears, hatch covers, compressor lubrication, turbo blowers, etc.
- Gear oils:
  These oils are prepared from highly refined base stocks containing sulfur-phosphorous additives to get anti-wear and extreme pressure characteristics. The oils have superior thermal stability, low foaming, excellent de-emulsibility, and rust protection characteristics. The oils are applied in the gear boxes where splash or force lubrication is used.
- Steam cylinder oils:
  These oils have excellent thermal and chemical stability. These oils are special quality mineral oils without additives, and it separates readily from exhaust

steam and condensate. The oils are used for worm gears operating under moderate load and speed.
- Stern tube oils:
  It is a compounded oil recommended for lubrication of stern tubes. These oils require separate Navy approval.

3.1.3 Lube Oil Marketing Specifications

While various OEM (original equipment manufacturers) in the automobile and industrial machinery equipment sectors built up equipment and machineries as well as doing continuous development in these sectors, they have to work in line with the lubricant manufacturing industries and keep pressures on the lubricant industry to meet the improved design of equipment and machineries.

To facilitate above, there are four major organizations who developed international standards for the lubricants as follows:

- API (American Petroleum Institute).
- ILSAC (International Lubricant Standards and Approval Committee).
- ACEA (Association of European Engines Manufacturers).
- JASO (Japan Automobiles Standards Organization).

API also made standards for base stocks as discussed in Chapter 2. They have made the standards for lubricants also in line with the standards for base stocks but in co-ordination with ILSAC, who basically developed and approved the lubricant standards for each category of automobiles and machines. So, there are basically three standards. The lubricant manufacturers, while taking the approval from OEM, simultaneously, have to conform to API / ILSAC standards or ACEA standards or JASO standards in respective countries where the standards are applicable to be followed. ACEA standards are followed by European countries; hence, vehicles made with European standards marketed or manufactured in non-European countries, have to come out with matching lubricant standards like API / ILSAC or JASO as followed by the countries to facilitate lubricant availability in those countries because lubricant manufacturers also follow the standards as set out by the Government body in those countries; similar be the case for the vehicles made with JASO standards. For example, in India, there is BIS specification, IS-13656-2002 for internal combustion engine's crankcase oil (lubricant) for gasoline as well as diesel fuel fired categories. In the specification, it is found that for each IS category there is matching international category like API / ILSAC, ACEA, JASO, etc.

The standards developed are divided into following sub-categories:

- Classification standards.
- Physico-chemical properties standards.
- Engine test performance standards.

The classification standards refer to scope and area of application.

Physico-chemical standards refer to all physical and chemical properties like density, viscosity, pour point, flash point, etc. as physical properties, and CCR (carbon conradson residue), sulfur content, ash content, metals content, oxidation stability, deposit formation, etc. as chemical properties.

Engine performance test /standards refer to bearing weight loss, piston skirt varnish, evidence of ring sticking or wear, piston ring and liner scuffing, oil consumption, kinematic viscosity increase/decrease during course of use in hours, average cylinder wear, piston cleanliness, etc.

API standards are divided into three categories, viz. API S, API C, and API F. API S series refers oil standards for gasoline engines; API C refers to commercial cars, i.e. diesel engine service; API F refers to heavy duty engines.

Similarly, ACEA standards are also sub-divided into categories called sequencing, viz. sequence A, B and C. Sequence A stands for gasoline engine, B stands for diesel engines and C stands for heavy trucks.

While, API / ILSAC, ACEA, JASO developed the standards, the testing methods of the properties as in the standards refer to ASTM standard methods even if some standards have their own testing methods.

In the physico-chemical properties standards, where viscosity of the lubricants is the most important property, the specifications are categorized in terms of viscosity which is also expressed in standard known as SAE standard where SAE stands for society of automotive engineers. For example, viscosity grade, 10W30 means it is multigrade lubricant where 10 stands for viscosity in winter and 30 stands viscosity in summer. However, in each viscosity grade, there are other physico-chemical properties which have been specified in the specification: the IS-13656-2002 as mentioned above.

In engine test performance standards, the above performance evaluations in engine test, as mentioned above, are again established to develop standard procedures, and the performance test standards so classified are named as CLRL-38, Sequence-IID, IIID, IIIE, VD, VE, VIA, CAT IH2, etc. which in detail are described in IS-13656-2002.

In addition to the above mentioned three categories of standards, the lubricant specification in IS:13656-2002[11] also provides a table, showing requirement of finished product identification and product conformation as lubricant, which also specifies physico-chemical properties not adequately covered as mentioned earlier above.

Similarly, for two-stroke engines also, the Indian specification, viz. IS-14234, covers all the aspects of international standards as discussed above. Many countries, like India have developed their own Govt. specifications for lubricant based on the API / ILSAC or ACEA or JASO standards.

Note: similarly, for two-wheelers also, there is IS specification[12] equivalent to API/ILSAC and ACEA/JASO.

3.1.4 Processes Followed in Lube Oil Production / Blending

Lube oil production is a blending process with base stock(s) as the main feed where various additives are mixed homogeneously in proportions as per formulations developed by the lube marketing companies to get the finished lubricants (lube oils) of different grades meant for use in particular categories of vehicles as discussed in detail above.

Blending is a physical process. Hence, lube blending is also a physical process but there may be some exothermicity / endothermicity while the additive components mix with all other components and base stock(s) in the blending vessel. Also, lube blending is a batch process and not a continuous process. In blending, homogeneous mixing among all components happen through diffusion process. Diffusion can be of three types, viz. molecular diffusion, vortex diffusion and convectional diffusion. To make it a perfect blending, the lube blending process is carried out such that all the above diffusion processes occur during blending activities. A residence time is provided to ensure molecular diffusion to occur; stirring is done to ensure vortex diffusion to occur, and heating is done to ensure convectional diffusion to occur.

Blending between base stocks or blending between base stock(s) and additive(s) means blending of the respective physical properties of the said components in the blend where all the components should be compatible to each other, i.e. they should integrally mix among themselves instead of dissolving one or two components by a third component thereby forming layer in the blend where at the same time the components should not chemically react among themselves.

For example, properties like density, viscosity, flash point, CCR, pour point, etc. and also the mechanical properties like anti-wear, skurfing, etc. contributed by each component, i.e. base stock(s) and additive(s) should have adding properties to be manifested in the blend, and hence, the name, 'additive', also justify.

Example of evaluating viscosity of the blend:

Alternative-1

- Measure viscosity of the blend.
- Change the blend ratio.
- Measure viscosity of the new blend.
- Repeat the exercise four to five times.
- Plot all the blend viscosity data as Y-axis vs. blend composition as X-axis.
- From the graph, with a particular blend composition, viscosity of the blend can be determined.

Note:

i. When there are more than two components in the blend, it is not possible to make two-dimensional graph to evaluate blend viscosity; in that case, it is better to collect at least ten sets of laboratory analysis data on viscosities of the blend for each set of blend composition, and then to develop a software simulation model, and test the simulation model for another fresh set of data, correct the simulation model, if required, and finally use the model for evaluating blend property while using base stock(s) and additives of known properties.

ii. Similar to above, each simulation model is to be developed for each blend property.

Alternative-2

- Measure viscosity of each component.
- Know the concentration of each component in the blend.
- For each viscosity, get the corresponding viscosity constant from the viscosity chart.
- Multiply the concentration with viscosity constant of each component; then sum up to get the viscosity constant for the blend.
- From the viscosity chart / Table 3.7, get the viscosity of the blend against a value of viscosity constant.

Note:

i. Similar procedures can be followed for other properties as mentioned above.

ii. For evaluation of properties like, anti-oxidant, anti-corrosion, detergency, dispersion, antifoam, etc. follow the Alternative-1 as proper representative blend tables may not be possible to be arrived in Alternative-2.

Even with all above in considerations, the manufacturer also should be concerned on the following in addition to the formulation developed by them:

- Base stock-additive interaction.
- Additive-additive interaction.

The above affect not only the quality of the finished lubricant but also in the blending process while making the lubricant.

Note: whether viscosity at 40°C or at 100°C, the values for all constituents in lube oil blending fall within the data range as given in Table 3.7.

Calculation procedure:

$$W_1 \times F(V_1) + W_2 \times F(V_2) = (W_1 + W_2) \times F(V)$$

Where W_1 and W_2 are the weights of the component one and two, and $(W_1 + W_2)$ is the weight of the blend; similarly, $F(V_1)$, $F(V_2)$ and $F(V)$ are the corresponding viscosity functions.

Illustration-1

70 MT of one component with viscosity of 120 cst. having viscosity function obtained as 33.76 from the Table 3.7 is to be blended with 30 MT of another component with viscosity of 50 cst. having viscosity function to be 30.86 as obtained from same Table 3.7.

TABLE 3.7
Chart on Viscosity vs. Viscosity Function

Viscosity, Cst	F(V)	Viscosity, cst	F(V)	Viscosity, cst	F(V)	Viscosity, Cst	F(V)	Viscosity, Cst	F(V)
0.6	4.86	1.6	9.04	2.6	13.91	3.6	16.69	70	32.04
0.7	2.15	1.7	9.7	2.7	14.25	3.7	16.91	75	32.27
0.8	0.00	1.8	10.31	2.8	14.58	3.8	17.12	80	32.48
0.9	1.76	1.9	10.88	2.9	14.88	3.9	17.32	85	32.68
1.0	3.25	2.0	11.4	3.0	15.18	4.0	17.52	90	32.87
1.1	4.53	2.1	11.89	3.1	15.46	4.2	17.89	95	33.05
1.2	5.65	2.2	12.34	3.2	15.72	4.4	18.24	110	33.49
1.3	6.64	2.3	12.77	3.3	15.98	4.6	18.57	120	33.76
1.4	7.52	2.4	13.17	3.4	16.23	4.8	18.88	130	34
1.5	8.32	2.5	13.55	3.5	16.46	5.0	19.17	140	34.21
5.4	19.72	6.8	21.25	8.5	22.63	11	24.11	150	34.41
5.8	20.21	7.0	21.44	9.0	22.97	12	24.58	160	34.6
6.2	20.65	7.5	21.87	9.5	23.28	13	25	170	34.77
6.4	20.86	8.0	22.27	10.0	23.58	14	25.38	180	34.93
15	25.73	22	27.55	34	29.39	46	30.56	190	35.08
16	26.05	24	27.93	36	29.62	48	30.71	200	35.22
17	26.85	26	28.28	38	29.83	50	30.86	220	35.48
18	26.62	28	28.59	40	30.03	55	31.2	240	35.71
19	26.88	30	28.88	42	30.21	60	31.51	260	35.92
20	27.11	32	29.15	44	30.39	65	31.79	280	36.11

The resultant viscosity function of the blend works out to:

$$70 \times 33.76 + 30 \times 30.86 = (70 + 30) \times F(V)$$

i.e. $F(V) = 32.89$.

Now from the same Table 3.7, viscosity corresponding viscosity function at 32.89 comes to be 90 cst.

Process description – a case study

The base stock(s) is / are pumped from the storage tank(s) to the blending vessel at a fixed rate controlled by a mass flow meter. After the desired quantity of the base stock(s) is / are taken up to certain level of the vessel, mild heating of the vessel content is done to ensure dehydration of the base stock(s) if there is moisture in the base stock(s). Then, the additive components dosing is also started by dosing pumps at the metered rates with the help of mass flow meters; while blending the additives into the base stock(s), continuous stirring of the vessel content is also carried out with help of an installed motorized agitator in the vessel. Again, heating up to a temperature of about 60–80°C is carried to ensure proper inter-mixing of all components in the blend. Heating is stopped after mixing process is over followed

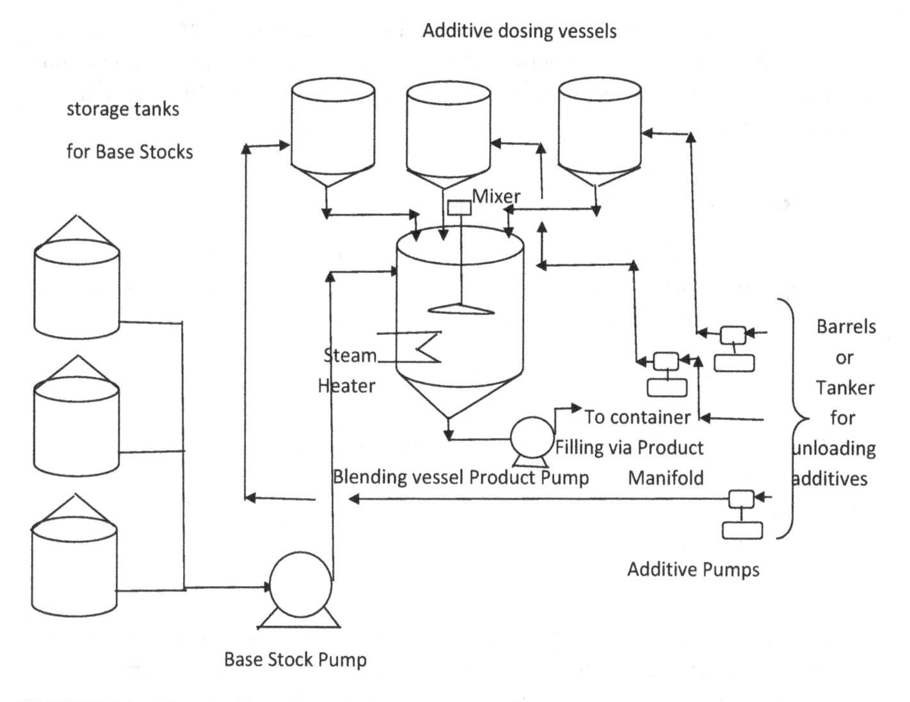

FIGURE 3.9 Case study schematic diagram of a lube oil blending plant.

by providing a residence time of a few minutes in the vessel to get the stability in the blend.

As in a lube blending plant of a commercial capacity, a large quantity of finished product along with a very large number of lube oil grades are produced, to ensure production efficiency at high by making continuity of operation, i.e. through implementation of automation in blending like carrying out the blending in a number of vessels with different time cycle such that there would be continuous flow of blended product through product pipe, and also to avoid contamination between the grades, a number of product manifolds are installed which are also operated through automation.

As such, the whole process is carried out with supervisory control from a remote-control room through a smart automation system.

A case study schematic flow diagram of lube blending process is shown in Figure 3.9.

Illustration-2

In production of lubricant of certain grade as per specification, it is found that desired qualities could not achieved though operating parameters of blending process have been followed. How, to overcome this problem?

Solution: both LOBS and additives qualities checked and found in order. Then additive formulation is marginally changed to get back the desired qualities of the product lubricant in the laboratory analysis; if the problem remains, then LOBS feed blend to be changed accordingly with ratio increase of the feed component having better qualities with respect to the properties which fail in the final product because additive formulation is not subject to major variation as the same is established after long duration R and D activities.

3.1.5 Branding and Marketing of Lube Oils

World lubricant consumption is about 30 MMTA (million metric ton per annum) with market capitalization of about 120 billion USD with CAGR of about 3.5% with seven major markets as below:

- USA
- Singapore
- Germany
- Japan
- China
- South Korea
- India.

Lubricant market is very competitive due to continuous development in automotive and industrial engines across the world requiring lube marketing companies also meet the objective of OEMs (original equipment manufacturers) of the vehicles and industrial machines. Also, unlike base stock manufacturing companies where the industry is capital intensive, lube oil blending plant and its operation requires very less capital and less operating cost except higher base stock price making it to be in the category of medium scale industry resulting in tough competition to sell the products.

Initially to meet stringent norms of lubricant set up API / ILSAC, ACEA, and JASO (internal organizations for standardization of lubricant), major corporate organizations in oil and gas sector like MOBIL (later on merged as Exxon-MOBIL after taking over of MOBIL by Exxon, USA), SHELL, BP, CALTEX, CASTROL, etc. were dominating the market, but subsequently Energy Inc, South Korea, Sinopec Lubricant, China, and Indian oil, India came into the fray and have occupied a major share of the lube market.

As lubricants are not directly consumed by the individual consumers and instead they purchase the lubricant based on the recommendation by the OEM of the vehicles / machines, reputation of the lubricant sellers plays a prominent role in the market instead of merely claiming meeting the specifications of the lubricant(s); here, comes the need of branding a product to be accepted by the people instead of going into the details of the product or even sometimes without giving much importance to its price.

In branding the product(s), all the above-mentioned marketing companies have necessarily incorporated a trade name to the particular grade of lube oil while mentioning compliances of necessary specifications and approvals. For example, some international brands are given as below:

- Castrol lubricants of different sub-grades (for gasoline and diesel engines).
- GS-Caltex lubricants by Caltex, UK.

- MOBIL lubricants by Exxon-MOBIL, USA.
- Servo-Super (SS) lubricants and servo-superior grades also by Indian Oil, India.
- S-OIL lubricants by S-OIL Corporation, South Korea.
- SYNOPEC lubricants by Sinopec Lubricant Co. Ltd. China.

A summary of brands of automobile lube oil of different marketing companies is given in Table 3.8.

Lubricant companies of respective countries are the market leader in their own country, and in the nearby countries where there is no indigenous lubricant company.

Even if there is competition among the lubricant marketing companies in a country on passenger automobile lube oil segment, the market in industrial lubricant is monopolized by the respective indigenous lube marketing companies as the industries in a country prefer not to import the lubricant as they require bulk quantities which are not only economic in purchasing indigenous lubricant but also inventory friendly; hence, they ask the OEMs to provide the indigenous equivalent list of lubricants for their machines.

3.2 GREASE

Grease is a semi-solid or sometimes solid form of lubricant as mentioned in sub-Section 3.1. Grease is a mixture of base oil and metallic soap added with various additives as discussed earlier. Though soap is generally defined as sodium salt of stearic acid, conceptually, it is basically a metallic salt of fatty acid. In solid form, soap has also got lubricity, i.e. viscosity, i.e. frictional resistance to flow while got softened in presence of a liquid, for example, in presence of water in cosmetic and laundry soaps, and in presence of base oil in grease.

Like synthetic lube oils or base oils, there are synthetic greases also developed for the lubricant industry, for example, molybdenum disulfide and graphite. Like semi-synthetic lube oils, there are semi-synthetic greases also like mineral base oil-based grease mixed with graphite or molybdenum disulfide.

Metallic soaps like lithium, calcium, aluminum, and / or barium stearate are used to make the modern greases of different grades. Titanium based grease also has been developed and patented by Indian Oil Corporation, India which has also become popular in the lubrication industry. Instead of metallic soaps, thickening agents, clays, PTFE powder, poly urea or graphite gel are also used in the grease blends.

3.2.1 Properties and Specifications of Greases

Like lube oils, greases also have the same properties like viscosity, VI, flash point, CCR, thermal stability, oxidation stability, anti-rust, anti-corrosion and anti-wear properties, dispersion, detergent properties, etc. except pour point property because grease, as mentioned above, is already in solid or semi solid form where this property is meaningless; hence, instead, drop point is a meaningful property for greases. Regarding antifoam property, it is known that soap should have good foaming property when mixed with water or oily water, and even in its neat form, there is some

TABLE 3.8
Brands of Automobile Lube Oils Marketed by International Lube Marketing Companies

Indian Oil	Castrol	MOBIL	GS CALTEX	SHELL	S OIL	SYNOPEC
1. SS 10W30 or 40 or 50, 20W 30 or 40 or 50 2. SS 5W-30, 10W-30, 10W-40, 20W-40, 20W-50	1. Castrol Active 2. Castrol Edge 3. Castrol GTX 4. Castrol Magnetic	MOBIL 1 1. 0W-20 2. 5W-20 3. 5W-30 4. 10W-30	1. Kixx PAO 1 2. Kixx G1 Synthetic 3. Kixx G1 Synthetic Power 4. Kixx Gold Supreme	1. HELIX ULTRA 5W-30 2. RIMULA 15W-40 3. HELIX HX6 Synthetic 5W-40	1. S-OIL 7 10W-30, 10W-40, 5W30 2. S-OIL 7 4T Scooter 10W-40 3. S-OIL 7 PAO 0W-40	1. JASTAR J700F Plus 5W-30 2. TULUX T700 CK4 15W-40 3. TULUX T500 C14 15W-40

water present in the soap, but in case of greases, which may get softened towards liquidity at high temperature during operation of the machine, there should not be foaming in grease preventing its lubricity effectiveness; hence, there must be some antifoam added in the grease formulation to prevent such foaming.

As the soaps don't have that much lubricant property, as mentioned above, to that extent which the base oils have, the greases thus made using soap in the blends can't have that much effectiveness in those properties; hence, greases are used comparatively in the low speed or low speed component of the machine, low load machines as compared to lube oils used in very high speed and high load machines.

Greases fall under the category of non-Newtonian fluids, and resemble the properties of Bingham Plastics unlike base oils which fall under category of Newtonian fluids, i.e. greases exhibit a complex rheological behavior unlike base oils.

In addition of common properties of base oils and lubricants, greases have some additional properties to be followed like:

- Drop point
- Viscosity-gravity constant (VGC)
- Aniline point
- Solubilizing property
- Penetration number.

Aniline point is representative of aromatic content which is also important in case base oils but that property was not discussed in Chapter 2 because there are other properties like VI discussed which is more important as increase in aromaticity decreases VI. Also, increase in aromaticity increases the value of CCR which has also been discussed in Chapter 2.

In respect to solubilizing property, aromaticity influences the same as aromatic compounds dissolve lighter oils to achieve homogeneity in the blend; however, higher aromaticity in greases is not desired due to disadvantage as mentioned above.

The three other important properties of the greases as mentioned above are discussed as follows.

- VGC:
  VGC is basically a multiplication of viscosity with gravity (density). Hence, the value of VGC not only represents viscosity but also gravity, i.e, how much soap is added in the grease blend on which gravity changes.
- Drop point:
  It is a temperature at which grease softens from semi-solid stage to liquid stage; the temperature at which first drop of liquid is formed is called drop point. This property is similar to drop melting point property of wax. The drop point of grease is important because during operation of the machine, temperature rises which may cause grease to soften to change into liquid stage; hence, lubricity properties of grease also would change because in semi-solid form, grease behaves like non-Newtonian fluid but in liquid state, it may behave as a Newtonian fluid; so, lubrication mechanism would change, particularly

shear resistance property; that's why shear reversibility characteristic property is very important in case of grease which means, though shear resistance would tend to decrease with rise in temperature due to structural change in the fluid, the grease regains its structural property during a course of time of operation resulting in reversibility of its shear resistance characteristics.

- Penetration number:

  It is representative of how soft is the grease. Its value is determined by dropping a standard cone like substance in a bed of grease sample, and the penetration of the cone inside the grease is measured in the scale of millimeter, and the value is expressed in 1/10th of millimeter, i.e. if penetration is 6 mm, the penetration number of the grease sample would be 60. The measurement philosophy is just like in bitumen sample but there instead of standard cone, standard needle is used for the measurement, and as per ASTM standard, measurement is done at 25°C after 60 strokes. The value of the penetration is called 'ASTM worked Penetration'.

  Against each penetration number, there would a fixed value of penetration index (like followed for 'bitumen') which can be used to evaluate the penetration number of a new blend of grease by changing the formulation.

  Based on the penetration number, gradation of grease has been done by NLGI (National Lubrication of Grease Institute), greases are graded from scale 0 to 6. For example, if the grease is very soft, it is called grade 0 grease; if soft, then called grade 1 grease, and if grease is very hard like solid, it called grade 6 grease. It may be mentioned that in 0 grade also, there are sub-classification, i.e. grade 0, 00, and 000. The sub-grade '000' is very thin grease closing to liquid stage, and thus it can't be filled in tubes.

  The consistency property of grease is manifested by its penetration number, i.e. if same penetration number is exhibited at different places of the surface of grease sample and the value is in conformity to meet the required all other properties of the grease, then it can be claimed that the grease is having consistency.

Important specifications of NLGI grades of grease are shown in Table 3.9.

Note:

i) The NLGI grades are based on worked ASTM penetration values only as shown above.
ii) Penetration number/index is measured as per ASTM D1403 and drop point as per ASTM D556.

3.2.2 Classification, Composition and Applications of Greases

Classifications of greases are the results of the following:

- Thickener used is metallic or non-metallic soap.
- Metallic soaps used are varying melting point.
- Using base oil of mineral base or semi-synthetic or synthetic nature.
- Neat synthetic grease without using thickener.

TABLE 3.9
Specifications of Typical Greases

NLGI grades	Thickeners	ASTM Penetration	Min. Drop point, °C	Uses
1	Sodium soap	310–340	180–185	Bearings and Gears
2	Do	265–295	180–185	Do
3	Do	220–250	180–185	Do
1	Aluminum soap	310–340	90	Gears
2	Do	265–295	90	Do
0	Calcium soap	355–385	80–90	Bearings, Hydraulic presses, Water pumps, etc.
1	Do	310–340	80–90	Do
2	Do	265–295	80–90	Do
3	Do	220–250	80–90	Do
2	Lithium-Calcium soap	265–295	165–180	Gears, Bearings, Water pumps, etc.
0	Lithium soap	355–385	170–190	Gears, Bearings, Joints, etc.
1	Do	310–340	170–190	Do
2	Do	265–295	170–190	Do
2	Lithium compound with synthetic base	265–295	>250	Aviation and Automobile gears and bearings.

Regarding using metallic soap of varying melting points, it is found that sodium soap melts at 150°C whereas calcium soap melts at 100°C; hence, sodium soaps are preferred than calcium soap as thickener.

It is the thickener which thus result in production of various grades of greases as shown in Table 3.9.

Accordingly various grades are classified, branded and marketed by the grease marketing companies according to their classifications as discussed below:

3.2.2.1 Classification

From marketing point of view, greases are classified into two categories:

- Automotive greases.
- Industrial greases.

3.2.2.2 Composition and Applications

There are various grades of greases as discussed above in the automotive class of greases. Various grease marketing companies have branded different grades accordingly, but all the categories fall under basic gradation of greases as done by NLGI which has their country chapter in all the countries. Accordingly, the marketing companies, in spite of branding their products as well claiming premium advantage of using their brands, have to meet NLGI specifications as obligatory.

- Automotive greases:

  This class of greases are used in accessories of automobiles where the speeds are lower or movement of the parts are intermediate or it be stationery parts requiring surface protection.

  Sub-classes:

  i. There is one class of greases, dark brown in color, NLGI grade 1 having drop point of 90°C. It is calcium base grease formulation with EP property and water repellence. It is used for general chassis lubrication including suspension and steering systems; also used for open and semi-enclosed gears, chain drives, etc. but it is not recommended for high temperature application due to its lower drop point.
  ii. This class of greases also are dark brown in color, sodium soap based, fall under NLGI grade 2 having drop point of about 170. These greases show excellent performance even at high temperature subjected to heavy shear. It is recommended for use in wheel bearings and various automotive grease applications however not exposed to moisture.
  iii. It is brown in color falling under NLGI grade 2 having drop point of 180, and is lithium based. It has high degree of resistance to oxidation, rusting and corrosion; it has good thermal and structural stability. It is multipurpose grease suitable for all automotive grease applications.
  iv. It is brown in color under NLGI grade 1 having drop point of 180, and is lithium-calcium based. It is a premium grade of grease having excellent water repellence, oxidation and structural stability along with higher anti-rust and anti-corrosion properties. It is used for lubrication of cartridge type taper roller bearings in Indian railways.
- Industrial greases:
  - General equipment and machinery greases:

    I. There are various brands under this category. The color of various subgrades varies from brown to dark brown with NLGI grade 2 and drop point at about 180 which have good water repellence, oxidation, thermal stability, anti-rust, and anti-corrosion properties and is used for both antifriction and plain bearing lubrication. In this performance category, there are exceptional formulations having NLGI grade 1, 0, and 00 which have excellent EP properties along with above mentioned properties, and these are used in automotive, earth moving equipment, gear couplings, electric motors, mining equipment and general industrial machineries.

    II. There are non-soap base class greases also in this category with NLGI grade 2 having higher drop point of 260. This grade has excellent water repellence, thermal stability even at high temperature. These are recommended for lubrication of machine elements, plain as well as antifriction bearings operating at high temperature. Greases in this grade can work in severe shock load conditions.

    III. There is light brown class of greases having NLGI grade of 2 and drop point of 90 which is lime based having good water repellence. This is used for plain and anti-friction bearing operating at temperature (−)18 to (+)60°C where ingress of moisture can't be avoided.

IV. With sodium base also, brands are available with light brown color having same NLGI grade but with drop point of 150. This class also shows good oxidation stability and anti-rust properties. This class of greases is used for screw down greasing systems of rolling mills, packed journal bearings and widely accepted in textile mills, machine tools, electric motors, etc. This grade is also recommended where heavy condensation of moisture and water ingress is unavoidable.

V. There are calcium-lead base grease brands available having color as black, NLGI grades 1, 0, and 00 with drop point of 204. Greases in this class also have good EP and anti-rust properties, and are used in anti-friction and plain bearings operating at medium speed under heavy/severe shock loads, and in gear couplings. These brands are also used in cane crushers, steel mills and mill roll of sugar factories.

VI. There are brands in the market which show good resistance to oil separation under pressure. These lithium complex greases. Greases in this grade have excellent mechanical stability and are used in engineering and marine industries.

- Chain greases:
  This class of greases fall under NLGI grade 6 having drop point of 90, i.e. very hard grease but with very low drop point made with formulation having anti-rust, anti-wear, and VII additives. These brands are recommended for lubrication of chains by lubricating the chains in the molten grease to facilitate narrow clearances between pin bush and the bush rollers.
- Graphited greases:
  In this class, there are brands of greases in NLGI grades 1, 2 and 3 with drop point of about 95 to 100, which are used for lubrication under high load and low relative displacement of inter-acting surfaces. These greases are recommended for leaf spring, hydraulic rams, plungers, slides, elevator cables, pantograph pans, for steel wire ropes, and in components with anti-seize purposes.
- Cement plant greases:
  In this class, the greases are grayish black falling under NLGI grades 0 and 00 with drop point of about 120, and are used as sprayable adhesive lubricant which have high quality colloidal graphite. These are applied in slow speed open gear drives, rotary kilns, cement mills, sintering plants, ball mills, etc.
- Lubrex cup greases:
  This class of greases has brands with NLGI grades 2 and 3 having lower drop point of about 90 due to use of calcium soap as thickener thus providing good water repellence. These greases are recommended for use in plain bearings operating under moderate to low temperatures where the churning and vibrations are within moderate limits.

4 Wax

Though the purpose of application of wax is different than lube oil or greases, wax falls under same subject category from point of view of feedstock used, i.e. all lube bearing crude oils are found to contain wax or even if some crude oils don't have lube potential due to not meeting the lube oil properties, they have wax potential of good quality. Also, the technology used for production of API grade-1 lube oils is also used for wax production, viz. in solvent dewaxing unit (SDU), during lube base oil production, lube base oil is the main product and wax is the by-product due to not meeting qualities of wax and hence being called a slack wax; but during wax production, wax is the main product and the oil is the by-product which can't be called lube oil due to not meeting the properties of base oil and hence it is called foots oil to be used as fuel in furnaces or as feedstock to hydrocracker or FCC unit. Thus, the operating conditions would be different in SDU during lube base oil production and during wax production; also, there would be some changes in the process and in using solvent just to optimize the objectives of the production. Accordingly, the full form of 'SDU' represents 'Solvent Dewaxing Unit' for lube base oil production, and it represents 'Solvent Deoiling Unit' for wax production purpose.

Wax can be of different categories. The above method of wax production comes under petroleum wax category but the wax in it comes under mineral wax category; similarly, waxes obtained from animal and vegetable sources come under natural wax category. On the other hand, if the wax is produced by chemical reactions within one or between two components, the wax obtained is called synthetic wax; for example, in the process of producing polyethylene from ethylene, the process can be controlled in such a way that polymerization can be terminated mid-way to get waxy ethylene, i.e. polymerization to stop after carbon number reaches to about 20; the wax formed should be called a by-product synthetic wax. But if the wax is produced starting with one or two single compounds reacting and polymerizing with objective to produce wax only as the main product, it is called fully synthetic wax.

From view point of feed sources, waxes then can be categorized as follows:

- Natural waxes:
    - Animal wax:
        - Beeswax, Chinese wax, Shellac wax, Spermaceti wax, Lanolin (wool wax).

DOI: 10.1201/9781003334422-4

- Vegetable wax:
  - Bayberry wax, Candelilla wax, Castor wax, Esparto wax, Japan wax, Ouricury wax, Rice bran wax, Soy wax.
- Mineral wax:
  - Petroleum wax:
    - Crude oil distillates, Crude oil tank bottoms.
  - Other mineral wax:
    - Ceresin wax, Montan wax, Ozocerite wax, Peat wax.
- Synthetic wax:
  - Polyethylene wax (PE wax).
  - Fischer Tropsch wax.
  - Chemically modified wax.
  - Substituted amide wax.
  - Polymerized alpha olefin wax (PAO wax).

4.1 PROPERTIES AND SIGNIFICANCE

Though wax is a solid product, it remains in plastic stage unlike metals, i.e. it can be bent by applying minor force. Like lube base oils, wax also has its own specific physico-chemical properties which should be met to use the wax for commercial purpose. The most important properties are discussed as follows:

- Oil content:
  Residual oil content should be minimum as per specification to keep it in solid form and prevent it from becoming soft or semi-solid during storage at high ambient temperature. The oil content is measured as %wt. of the wax sample.
- Drop melting point:
  It is the temperature at which the wax sample initiates melting, i.e. when first drop of melt forms. It is generally measured as °C. If oil content of the wax is higher, its drop melting point (DMP) would be lower and vice versa. Even with low oil content of wax, its DMP may be lower which is due to the crystalline structure of wax which however is benchmarked in such cases and incorporated in the specifications; this structure of wax has been discussed separately in the 'classification of wax' as follows.
- Color:
  It is also a very important property which restricts the use of wax in line with the objective of applications. As such, the color of the wax should be snow white to use it in the retail sector particularly when to use it as coat in food packaging. The color is an indirect indication of oil content of wax, viz. if there is oil content increase in the wax, its color would be dark and if it is negligible, the color of the wax can be as white as snow. The color is measured in ASTM scale or Lovibond scale of measurement, i.e. when color of the wax falls in ASTM scale, then even if its value is 0.5 ASTM (lowest in ASTM scale), wax color may be yellowish white, but when its color comes under Lovibond scale, say, Lovibond 25, the wax color would be snow white.

- Viscosity:
  When paraffin wax is used as candle, this property is not important as candle is in solid form, but when paraffin is used in blend to produce liquid derivative as final product required for particular application(s), then the viscosity of paraffin wax is important though it is not a criterion in standard specifications. The viscosity of paraffin wax in liquid form generally varies from 3.5 to 4.5cst. @100°C. For MCW, it is known that this wax is not brittle but very soft unlike paraffin wax; it is so soft that when one rubs his/her finger on the wax surface, a layer of wax gets stuck to his/her finger unlike paraffin wax. Hence, this wax sometimes is used as neat in addition to using it in the liquid blend, and thus its viscosity property is even more important than in paraffin wax. Its viscosity generally varies from 5 to 9cst. @100°C.
- FDA test:
  FDA full form is Foods and Drug Administration, USA which has set benchmark specification to use the wax in food packaging, i.e. to approve the wax quality as food grade quality or otherwise to use it for industrial purpose as non-food grade quality.

  If there is any oil more than 0.5%wt. in wax or even with such low oil content, if there is some carcinogenicity in the oil present in the wax, it does not pass the FDA test and the wax can't be used as food grade wax.

  FDA test is an UV absorption test measured as UV absorbance per centimeter path length. In this evaluation of the wax sample, the value of UV absorbance is measured at different wave length ranges in millimicron scale provided in the specification. The wax sample meeting these values can be called a food grade wax. The values are shown in the specification of wax as follows.

4.2 CLASSIFICATION OF WAX

From marketing point of view, wax is classified according to its structure and application. The molecular structures of waxes have been discussed in Chapter 1. Accordingly, as mentioned there also, wax is classified as follows:

- Paraffin wax.
- Micro-crystalline wax, commonly called MCW.

With respect to wax crystal morphology, there are two varieties, viz. platelets and needle type. Even in the needle type, there may be smaller fine needle like structures also.

The crystal structure of paraffin wax is platelet type or sometimes it may be agglomerates of big sizes needle structure also; but MCW crystal structures are found to be of smaller needles type. The properties and application of these two types of wax are as follows:

- Paraffin wax: its oil content remains on the lower side as compared to MCW and varies between 0.5%wt. up to 3.5%wt. categorizing the paraffin wax as type-I, type-II and type-III with type-1 with lowest oil content and type-III with highest oil content as detailed in the specification as follows. The drop melting point (DMP) of the paraffin wax varies from 45°C to 75°C but typical value is about 55°C for paraffin wax with oil content of 0.5%wt. Their lower DMP as compared to MCW is due to major concentration of straight chain paraffins with overall lower molecular weight of about 350 to 450 as compared to MCW.

The paraffin wax produced in SDU is generally processed in hydro-finishing unit where not only asphaltene, if any, is removed from the wax to get the desired color of the wax, but also, the traces of oil which remains in the finished wax gets hydrogenated thereby removing the carcinogenetic property of the wax which results in getting food grade quality of wax.

The paraffin wax with wax crystal structure of platelets type or mix with some big needle structures, generally observed to be hard in compressive strength but brittle in nature, for example, a candle which is a paraffin wax which is very hard but breaks into pieces when it falls on the ground from a height.

- MCW: its oil content is on the higher side as compared to paraffin wax and generally varies from 1.5%wt. up to 4.0%wt. thereby categorizing into different sub classes, viz. type-A, type-B and type-C. MCW have higher drop melting point (DMP) than paraffin wax and varies from 65°C up to 85°C with higher DMP means lower oil content and to be hardest one except type-E with higher DMP even at higher oil content. MCW with higher oil content, i.e. with lower DMP like 70°C, i.e. type-A grade is called semi-MCW and don't attract high market value unlike other grades.

  MCW due to presence of needle structure, that too, with finer needle structure, the crystal sizes become very tiny and thus called micro-crystalline. But, interestingly, MCW has higher DMP than paraffin wax as mentioned earlier even with higher oil content in the wax due to the presence of more naphthenic waxes, resulting in high molecular weight at about 600–800. The wax though has higher DMP but found to be very soft unlike paraffin wax, i.e. its ductility is higher as compared to paraffin wax. This soft structure makes the wax non-brittle and can be thus used in cosmetic industry if it is made up to food grade quality, otherwise also, it is very useful to be used as underground cable coating where the wax not becoming brittle prevent moisture ingress into the cable.

4.3 MARKETING SPECIFICATIONS OF WAX

Though waxes of different classes are free trade products, many countries have benchmarked their properties to facilitate both consumers and also producing and marketing companies come to a common platform in this area. Besides this, some reputed international marketing companies have branded their wax product by issuing separate set of specifications in order to facilitate buying their product instead from other manufacturing companies; these specifications also are found to be in line with those incorporated by many countries as mentioned above except there are marginally better values in one or two parameters in the specifications.

As examples, BIS specifications as Indian standards, and standards set out by one international marketing company, viz. Exxon-Mobil for paraffin wax and IGI (The International Group, Inc.) Global for micro-crystalline wax are reproduced as follows which demonstrate the importance of various properties of waxes as discussed earlier.

BIS specifications for paraffin wax are shown in Table 4.1.

Exxon-Mobil specifications for candle wax and semi-refined paraffin wax are shown in Table 4.2A and 4.2B respectively. Like paraffin wax, as shown in Table 4.2A, similar specifications for other grades of candle wax, viz. PARVAN-1310, 1330, 1341, 1380, 1381, 1402, 1420, 1451, 1470, 1471, 1520, 1540, 1580 are also available as set out by Exxon-Mobil.

The BIS specifications for MCW and IGI Global specifications for some branded varieties of MCW are shown in Table 4.3 and 4.4 respectively.

4.4 APPLICATION OF WAX

Application of waxes are decided based on classification, viz. paraffin wax or micro-crystalline wax, and accordingly the waxes find their usage in the industries as follows:

- Packaging industry:
  In this industry, food grade paraffin wax is widely used for coating of paper utilized in packaging of various food items. It is also used as paper cups and plates. The purpose of coating is to provide a non-porous cover on the base materials which are porous in nature; this coating can prevent exit of moisture from inside and also prevent entry of micro-organism from outside; it also provides an extra stiffness when used as coating to paper. Type-1 paraffin wax of specification-IS:4654-1993 or other equivalent internationally branded waxes are used in this sector, and the wax is colloquially called wax-paper.

  Though now-a-days instead of wax coating, plastic coatings are given, but these are discouraged from hygienic point of view.
- Pharmaceutical industry:
  Here paraffin waxes are used as additive for manufacturing of various pharmaceutical products, like balm, skin ointment, pharmaceutical cream, etc. as an active reagent to the base material. The waxes act as binding material and provide extra smoothness which aids in the uniform distribution of the ointment / cream on the skin.

  Both type-1 and type-2 paraffin wax of specification-IS:4654-1993 or other equivalent internationally branded waxes find application in this sector. Natural waxes like Beeswax and Carnauba wax are also used in this industry.
- Cosmetics industry:
  Paraffin waxes are also used in the cosmetics industry for preparation of cold creams, moisturizer creams and lotions as an additive along with other additives to the base material as per product formulation. But micro-crystalline wax (MCW) is more preferred in this industry due to its properties of modifying the rheology and smoothness of the creams.

TABLE 4.1
BIS Specifications of Paraffin Wax: IS 4654: 1993

Sl. no.	Characteristics	Type-1	Type-2	Type-2A	Type-3	Method of test Annexure & page of IS 1448	
1	Drop melting point, °C	45–75	45–75	58 min.	45–75	A	---
2	Ash content, %wt., max.	0.03	0.03	0.03	0.03		–4
3	Acidity, mg KOH/gm, max						
	-organic	0.1	0.1	0.1	0.01	-	2
	-inorganic	0.02	0.02	0.02	0.02	-	-
4	Saponification value, max	1.0	1.0	1.0	1.0	–	55
5	Oil content, %wt., max.	0.5	0.5	1.5	3.5	B	-
6	Color						
	-Lovibond in 18” cell, max	1.0Y	2.0Y	2.0Y	–	A	13
	-ASTM, max.	–	–	–	2.0	A	12
7	Acid discoloration test	Pass	–	–	–	C	–
8	UV absorbance per centimeter path length:						
	-280–289 millimicrons	0.15 max.	–	–	–	D	–
	-290–299 millimicrons	0.12 max.					
	-300–359 millimicrons	0.08 max.					
	-360–400 millimicrons	0.02 max.					
9	Odour, max.	1	–	–	–	E	–

Note: When ASTM colorimeter is not available, the color may be determined by Lovibond Tintometer in accordance with method A in p:13 of IS-1448 using ¼” cell. The corresponding color requirements are: (–)4.5Y + 0.4R, max.

TABLE 4.2A
Exxon-Mobil Specifications of Paraffin Wax for Candle Application (Branded as 'Parvan')

Product name ASTM test method (a)	DMP, °C Min/Max. D87	DMP, °F Min/Max D87	DMP, °F Typical D87	Oil content, %wt., Max. D721	KV, cst. @100°C typical D445	Needle penetration, 0.1 mm, typical @ 25°C/77°F 40°C/ 104°F D1321	SayboltColor, Min. D156-D6045	Flash Point COC, °F, Min D92
PARVAN-1270	51.7–53.3	125–128	126	0.5	3.4	16 108	+28	204 (400)
PARVAN-1290	52.8–54.4	127–130	129	0.5	3.5	16 110	+28	204 (400)

TABLE 4.2B
Exxon-Mobil Specifications of Semi Refined Paraffin Wax (Branded as 'Waxrex')

Product name ASTM test method (a)	DMP, °C Min/Max. D87	DMP, °F Min/Max D87	DMP, °F Typical D87	Oil content, %wt., Max. / Typical D721	KV, cst. @100°C typical D445	Needle penetration, 0.1 mm, typical @ 25°C/77°F, 40°C / 104°F D1321	SayboltColor, Min. D156-D6045	Flash Point COC, °C / °F, Min D92
WAXREX-1270	52–57	125.6–134.6	127	2.5 / 2	3.7	55 (Max.)	+25	200 (392)
WAXREX-1280	51.6–53.6	124.9–128.5	128	5–7 / 5.8	3.7	36	+28	204 (400)
WAXREX-1281	51–56	123.8–132.8	128	12 / 7	3.7	54	+28	204 (400)

TABLE 4.3
BIS Specifications of Micro-crystalline Wax (MCW): IS 13833: 1993

Sl. no.	Characteristics	Type-A	Type-B	Type-C	Type-D	Type-E	Method of test Annexure & page of ex. IS 1448	
1	Drop melting point, °C	65–70	70–80	70–80	85–95	85–95	A	--
2	Ash content, %wt., max.	0.03	0.03	0.03	0.03	0.03	–	4
3	Acidity, mg KOH/gm, max							
	-organic	0.1	0.1	0.1	0.1	0.1	-	2
	-inorganic	0.02	0.02	0.02	0.02	0.03	-	-
4	Saponification value, max	1.0	1.0	1.0	1.0	1.0	–	55
5	Oil content, %wt., max.	4.0	4.0	2.0	1.5	4.0	B of IS-4564	---
6	Color							
	-ASTM, max.	2.0	2.0	0.5	1.0	5.0	A	12
7	Needle penetration, 0.1 mm (25°C, 5 sec., 100g)	10–25	20–35	25–35	2–10	2–10	--	93
8	UV absorbance, max. per centimeter path length:							
	-280–289 millimicrons	0.15	0.15	0.15	0.15	–	D of IS-4654	---
	-290–299 millimicrons	0.12	0.12	0.12	0.12	–		
	-300–359 millimicrons	0.08	0.08	0.08	0.08	–		
	-360–400 millimicrons	0.02	0.02	0.02	0.02	–		
9	Odor, max.	0	0	0	0	0	E of IS-4654	---
10	Flash point, COC, °C, Min.	260	260	260	260	260	---	69

TABLE 4.4
Specifications of MCW of IGI Global

Type	Microsere* Grade	Melting Point, °F/°C Astm D127	Hardness, @77° F/25°C, Dmm Astm, D1321	Color Astm D1500	Application
Type 1- Laminating	5788A	125–145/52–63	25–45	<0.5	Adhesive, Candle,
	5715A	165–177/74–81	25–35	+16 (saybolt)	Cosmetic, Explosive,
	5799A	165–177/74–81	20–30	<1.5	Packaging.
Type 2- Coating	5801A	177–188/80–87	9–16	<0.5	Adhesive, Casting
	5806A	175–188/79–87	15–22	<2.5	Chewing gum, Ink
	5871A	175–188/79–87	15–22	<0.5	Packaging, Plastics,
	5890A	175–188/79–87	15–22	>16 (saybolt)	Rubber.
Type 3- Hardening	5999A	192–200/89–93	<10	<1.5	Adhesive, Casting
	5909A	192–200/89–93	<10	<0.5	Chewing gum, Ink
	5910A	192–204/89–96	<10	>16 (saybolt)	Specialty.

Note: The symbol * denotes registered trade mark.

When paraffin waxes are used, it should be food grade only with preference to type-1 or type-2A grade, and when MCW is used, it should be food grade only, i.e. type-A, B, C, or D only as per IS:18833-1993 specifications or equivalent internationally branded specifications.

- Tarpaulin industry:
  Paraffin wax of grade type-1 or 2 as per IS:4654-1993 specification or equivalent internationally branded specifications are used in this sector to provide water-proofing and water repellant properties to the tarpaulins produced. In the process, waxes are mixed with other chemicals and certain fillers and spread over tarpaulin cloth in a uniform manner.
  Natural waxes like Beeswax, Carnauba wax are also used in this sector and even plastic coating like Raffia and HDPE are also used now-a-days.
- Candle industry:
  Type-2 paraffin wax as per IS:4654-1993 or equivalent internationally branded waxes are preferred for candle manufacturing. According to the market data, about 70–75% of total waxes are consumed by this industry, and about 80% of waxes consumed are type-2 grade. The purpose of using the wax in the candle industry is to provide desired properties like strength and stiffness to the candle stem, slow sustained burning due to higher drop melting point of the wax, and bright appearance with better finish.
- PVC pipes and fittings industry:
  Type-2 paraffin wax and sometimes type-1 grade as per IS:4654-1993 or equivalent internationally branded varieties are used for coating purposes during the manufacturing of PVC pipes with lubricant property. They also provide certain protection against the wear and tear of the PVC pipes.

- Wax match industry:
  Type-3 paraffin wax as per IS:4654-1993 or equivalent branded waxes are adequate with respect to quality to use in this industry, but sometimes type-1 or 2 grades are used to make premium quality matches. Paraffin wax here is used as a primary raw material for manufacturing the sticks of the matches and as a paper coating in certain other matches. The primary advantage of paraffin wax is to increase the burning time and strength of the match sticks.
- Explosive industry:
  Grade-1 and Grade-2 paraffin waxes as per IS:7401-1987 or equivalent branded specifications are used as a water-proofing agent for paper shells used for packaging explosives and pyrotechnical components. They also find usage as additive in explosive formulations.
- Petroleum jelly industry:
  Most commonly, instead of using neat paraffin wax, a blend of paraffin wax and MCW is used as major additives along with several other heavier paraffin products in varying proportions for the preparation of petroleum jelly. The jelly is then utilized as a base material and additive for manufacture of several skin creams, ointments, balms, etc. Paraffin wax of type-1 and 2 grades and MCW of type-C and D grades as per specifications, IS:4654-1993 and IS:13833-1993 respectively or equivalent branded qualities are used in the blend.
- Tyre industry:
  MCW is most widely used in this industry as an additive to the rubber in the tyre making process. The addition of MCW to the rubber improves the flow of the product and also acts as anti-ozonate component for the protection of rubber from ozone.

  Some manufacturers have a tendency to use paraffin in place of MCW due to easy availability and less price of paraffin wax but that should not be a right practice.
- Rubber components and rubber products industry:
  MCW is widely used as an additive to rubber for manufacture of various products like pipes, belts, hoses, etc. with the same objectives as mentioned for the tyre industry.
- Paints and polish industry:
  MCW is used as an additive gellant during manufacture of various kinds of polishes, due to its properties of modifying the rheology and enhancing the gloss of the final product.

4.5 MANUFACTURING TECHNOLOGIES FOR WAX

Processes followed for recovery of mineral waxes present in petroleum sources are only discussed as this sector contributes in large quantity production of wax as compared to synthetic wax which is produced in small quantities in the industry. Various processes used for the production of paraffin and micro-crystalline wax are discussed as follows:

4.5.1 Solvent Deoiling of Vacuum Distillate

The crude oils from some specific oil fields are found to have very good wax potential; as such, the medium and heavy vacuum distillates of this crude oils have about 25 to 30% of paraffin wax content except the light vacuum distillates which contain lower wax content; also, the best quality paraffin wax is achieved by processing the medium vacuum distillates which yield about 25%wt. of paraffin wax while the quality of wax is not homogeneous by processing the heavy vacuum distillate. In the heavy vacuum distillate, there is a mix of paraffin and micro-crystalline wax, i.e. on crystallization of this distillate, there is non-homogeneity in crystal agglomerates due to formation of fine needle crystals along with platelets and bigger needle crystals. However, with fine-cut distillation in vacuum distillation unit, one can get the heavier narrow cut heavy distillate which on crystallization generates micro-crystalline wax (MCW), but the best way to get quality MCW is to use de-asphalted oil (DAO) as feedstock; the DAO production process has been discussed in Chapter 2. The presence of naphthenic wax, i.e. paraffin in the side chains of naphthene or poly-naphthene in the feedstock contribute to the generation of MCW on crystallization.

In some waxy crude oils, it is found that there is less concentration of aromatics as compared to Middle East crude oils where though wax potential is there but they have high concentration of aromatics and asphaltene in the vacuum distillates. Moreover, as there is much less concentration of aromatics or asphaltene in the medium vacuum distillates as compared to heavy vacuum distillates in such special waxy crude oils, there is no need for processing this medium vacuum distillate through aromatic extraction unit, and as such direct solvent deoiling of the medium vacuum distillate can be done to get paraffin wax, and thus the process of wax recovery becomes economically attractive.

- Principles of solvent deoiling:
  - Deoiling temperature:

  It is same as solvent dewaxing process as discussed in Chapter 2 except in there, the objective was to get base oil as the main product after removing the wax from it, and the wax so obtained is called slack wax as a by-product, but in solvent deoiling, it is vice versa, i.e. objective is to get the wax as the main product and the oil which is separated out from the wax is the by-product.

  The abbreviated name of this unit also is 'SDU' but it represents 'solvent deoilinng unit' instead of 'solvent dewaxing unit'. In keeping such objective as mentioned above, while the quality of the product, wax would be good, i.e. with very negligible oil content in it, the quality of the by-product, i.e. oil would be left with contamination of wax; in other way, it can be said that in solvent deoiling unit, to get the wax as a product with very negligible oil content, operating conditions, particularly, the final chiller outlet and filtration temperature is kept on the higher side (about +3 to +5°C) unlike in solvent dewaxing process so that the oil doesn't crystallize and remains in the liquid form thereby helping it to be filtered out; but in doing so, some lighter wax molecules would also

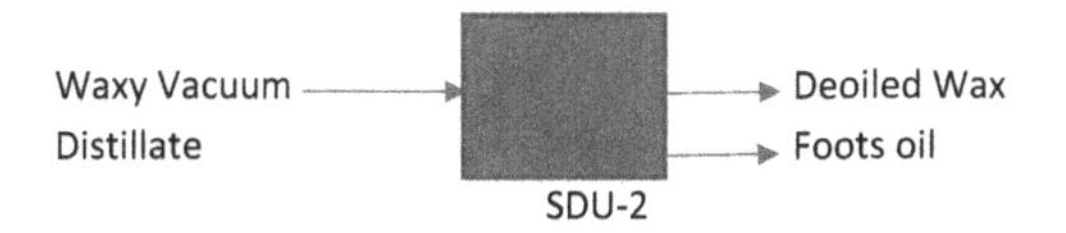

FIGURE 4.1 Solvent deoiling of vacuum distillate.

remain in the liquid form and get filtered out along the oil. The by-product, oil being contaminated with a trail of lighter wax is called 'Foots Oil'.

This direct deoiling of the distillate is also preferred as it is found that the foots oil has not enough lubricity to use this as lube oil, and thus it is either used as fuel in the furnaces or as feedstock in hydrocracker or FCC unit.

The above discussed process in deoiling is represented by a block flow diagram (BFD), Figure 4.1.

In the BFD, as shown in Figure 4.1, when the feed is light vacuum distillate or medium vacuum distillate, the wax obtained would of paraffin wax grades; but when the feedstock is narrow cut heavy vacuum distillate or more preferably DAO (deasphalted oil as discussed in Chapter 2), the wax obtained would be of microcrystalline grades.

- Selection of solvent, miscibility temperature and solvent dilution:

In Chapter 2, sub-Section 2.4.1.3, the subject has been discussed for processing of aromatic extracted raffinate in SDU (solvent dewaxing unit). There the use of dual solvent (MEK + toluene) was the objective to get the quality oil (with minimum ingress of wax in it) that too where quantity of oil in feed is much more than wax content in it, some solvent like toluene was needed with good solvent power to dissolve the oil into it and at the same time, some other solvent like MEK with very good anti-solvent property to wax even if it has less solvent power to oil unlike toluene. But in solvent deoiling unit (SDU-2), the objective is not to get quality dewaxed oil but to get quality deoiled wax; hence, the alternative solvent like MIBK (methyl-iso-butyl ketone) has been selected where MIBK has good solvency to oil (marginally less than toluene) but at the same time has good anti-solvent power to wax. But anti-solvent power of MIBK is marginally less than that of MEK. But when some dissolution of wax into oil phase can be accepted in order to get deoiled wax without or with negligible oil content, it is preferred to select a single solvent like MIBK to get the solvent recovery circuit in more simple form than while using a dual solvent system.

Also, the total solvent dilution to the feed is not higher in case of using MIBK as only solvent as against using dual solvent (MEK + toluene) as well as miscibility temperatures with MIBK as solvent and with MEK + toluene as solvent are also nearly same rather slightly lower with MIBK as solvent; thus solvent-feed mix heating temperature can be maintained at the same level of about 65–70°C before sending it to water cooler to initiate wax crystal nucleation at cooler outlet.

TABLE 4.5
Properties Comparison between MIBK, MEK and Toluene

Solvent	Mol. Wt.	Boiling point (B.P.), °C	Density, D_4^{20}	Abs. Visc., C_p at 20°C	Flash Point, °F (Closed cup)	Freeze point, °C	Sp. Heat, Liq., @10°C, Cal/gm	Latent Heat of vap. at B.P., Cal/gm
MIBK	100	115.9	0.802	0.585	60	−84.7	0.460	86.5
MEK	72	79.6	0.805	0.425	19	−86.7	0.498	106
TOL	92	110.6	0.867	0.587	40	−95	0.392	86.0

The basic advantage of using wet MIBK is its relative immiscibility with water as compared to MEK; at typical condensing temperature in solvent recovery circuit where from the MIBK-water condenser accumulator vessel, MIBK rich-negligible water phase can be directed to solvent stream as dilution solvent, the water rich phase which contains very negligible concentration of MIBK (max. 2%wt.) can be drained out without processing through azeotrope distillation column unlike in MEK+toluene solvent system.

Another advantage of using MIBK as solvent is its lower latent heat of vaporization as compared to dual solvent, MEK + toluene which helps in saving heat energy consumption in the process.

The comparative physical properties of MIBK, MEK and toluene are shown in Table 4.5.

- **Refrigeration/crystallization – a case study report:**
  While similar temperature profiles and solvent dilution ratios are maintained in the refrigeration/crystallization section, the number of scrapper chillers in the series can be reduced to two against three in solvent dewaxing as final chiller outlet temperature required at about +3 to +5°C as compared to (−)14 to (−)17°C in solvent dewaxing; thus, there is no chance of chocking of final scrapper chiller due to icing of water if it ingresses into the solvent.

  However, the outlet temperature of the solvent chiller used to chill the solvent (in order to use the chilled solvent as cold wash spray in the filters) though higher than in solvent dewaxing but still it remains in sub-zero temperature, i.e. (−)3 to (−)5°C; hence the final solvent chiller may get choked with ingress of water, if any, in the solvent. To avoid such problem, some refineries, instead of using direct refrigeration in the chillers, use another chilling medium like 30% water solution of ethylene glycol which first gets chilled in the refrigeration circuit and then it moves to all chillers for subsequent chilling of feed-solvent mix streams as well as solvent streams; the water in ethylene-glycol solution doesn't congeal at that temperature, and also it can be used in tube side of solvent thus facilitating the flow of solvent through shell side of the shell and tube chiller; hence, with cold solvent flowing in the shell side, there is no chance of cold solvent flow choking and cold wash flow to filters can be maintained without interruption except there may be chocking of cold wash spray nozzles in the filters but that can be de-chocked quickly by isolating it for sometimes and taking alternate filter in line for the time being or alternatively flowing a stream of hot solvent through the cold wash nozzles for 5 minutes approximately.

Due to the above advantage, the use of solvent drying column circuit as shown in Chapter 2 sub-Section 2.4.1.3 can be avoided in solvent deoiling process.

- **Filtration – a case study report:**

The chilled feed-solvent mix from refrigeration / crystallization section enters a series of rotary vacuum filter kept in parallel like in SDU discussed in Chapter 2. the operation of the filters is guided by the principle as shown through following schematic diagram of the filter as in Figure 4.2.

The bottom half of the horizontal cylindrical vessel of each filter is called 'Vat' where the chilled feed-solvent mass enters; the upper half of the vessel is called 'Hood' which acts as inert gas blanketing space which is connected to an inert gas line; a rotary cylindrical drum (as shown in Figure 4.2) is placed inside this closed cylindrical vessel keeping a space clearance from the vat for free rotation of the filter drum and facilitating a level of liquid pool of chilled feed-solvent mass in the vat. In the rotating drum inside, there are lots of small pipe for collecting the filtrate under vacuum created by a vacuum compressor

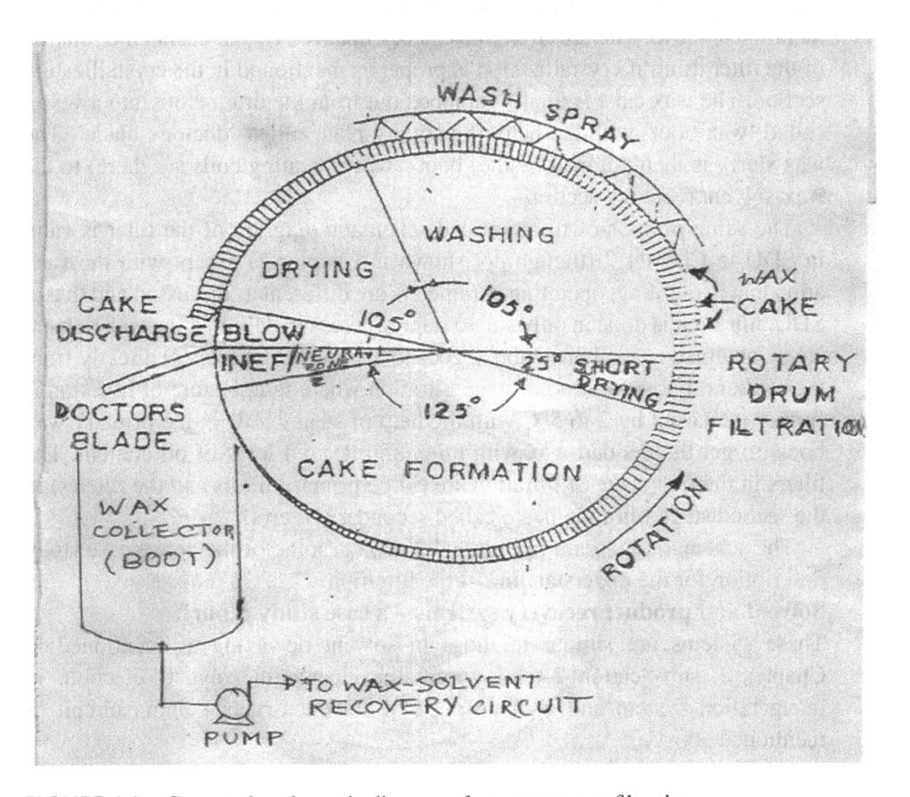

FIGURE 4.2 Case study schematic diagram of rotary vacuum filtration.

connected to the filtrate vessel at its downstream with filtrate vessel connected to rotary drum outlet where all the pipes as mentioned above (called lead-lag pipes) meet to a stationary plate connected with three filtrate pipes leading to the filtrate vessel. A specially designed plate called 'bridge plate' with many holes is connected to these lead-lag pipes, and this bridge plate in rotating condition in turn is connected with the above mentioned three stationary filtrate pipes with sealing mechanism to avoid any short circuit of filtrate into the vat. The cylindrical surface of the rotary drum is covered with micro-porosity filter cloth wrapped homogeneously by steel ropes to keep the cloth in tight-hold with the rotary drum. The filtrate from the bottom part sucked through the lead-lag pipes during drum rotation into bottom pipe of these three stationary pipes is under highest vacuum and is called bottom filtrate; the drum while rotating upwards in or above the submerged pool of liquid mass in the vat sucks further filtrate through lead-lag pipes and collects into middle pipe of these three stationary pipes and this filtrate is called middle vacuum filtrate; the drum while moving further upwards completely leaving the liquid mass of the vat continues to suck further filtrate through lead-lag pipes into the third stationary filtrate pipe and is called top vacuum filtrate. To facilitate maximum removal of oil from the wax cake formed on the filter cloth, a cold solvent is showered to the top of the drum through various pipes fitted with spray nozzles when this cold solvent along with residual oil in the wax cake is sucked into to the third pipe as mentioned above. The filtration rate can be enhanced by increasing the rotation of the filter drum if crystallization is proper as mentioned in the crystallization section. The wax cake is finally scrapped out from the drum cloth into a vessel called 'wax boot' with the help of a sliding plate called 'doctor's blade'. The wax slurry is then pumped by the 'boot' (where heating coils are there) to the wax-solvent recovery section.

The same process of filtration and schematic diagram of the filter is valid in SDU in Chapter 2 (though not shown in Chapter 2) except with the variation that in deoiling, operating parameters are different as discussed and that in SDU, filtration is done in only single stage whereas in deoiling, the wax-solvent slurry obtained as explained above goes to another set of filter(s) directly from these filters' boots for second stage filtration where temperature of filtration is further increased by 2 to 5°C with the help of steam coils in the primary wax boots to get the deoiled wax with minimum desired level of oil content. The filters in the first stage of filtration are called primary filters and the filter(s) in the second stage filtration is/are called secondary filter(s).

The schematic diagram in Figure 4.2 depicts some of the above case study description for the understanding of the filtration.

- **Solvent and product recovery systems – a case study report:**
These systems are similar to those in solvent dewaxing as mentioned in Chapter 2, sub-Section 2.4.1.3 except the change in solvent selection, in refrigeration system and with no need of solvent drying column circuit as mentioned above.

- **Case study process description and process flow diagram:**
  The process flow description and process flow diagrams are also same as in sub-Section 2.4.1.3 of Chapter 2 with all the exceptions as mentioned above. However, the modified process flow diagram for solvent deoiling unit, SDU-2 (as mentioned above) is shown as Figure 4.3.
- **Case study operating conditions in solvent deoiling:**
  As explained, the typical operation conditions of relevant operating parameters in solvent deoiling unit are shown in Table 4.6 and 4.7 for paraffin wax and MCW respectively.

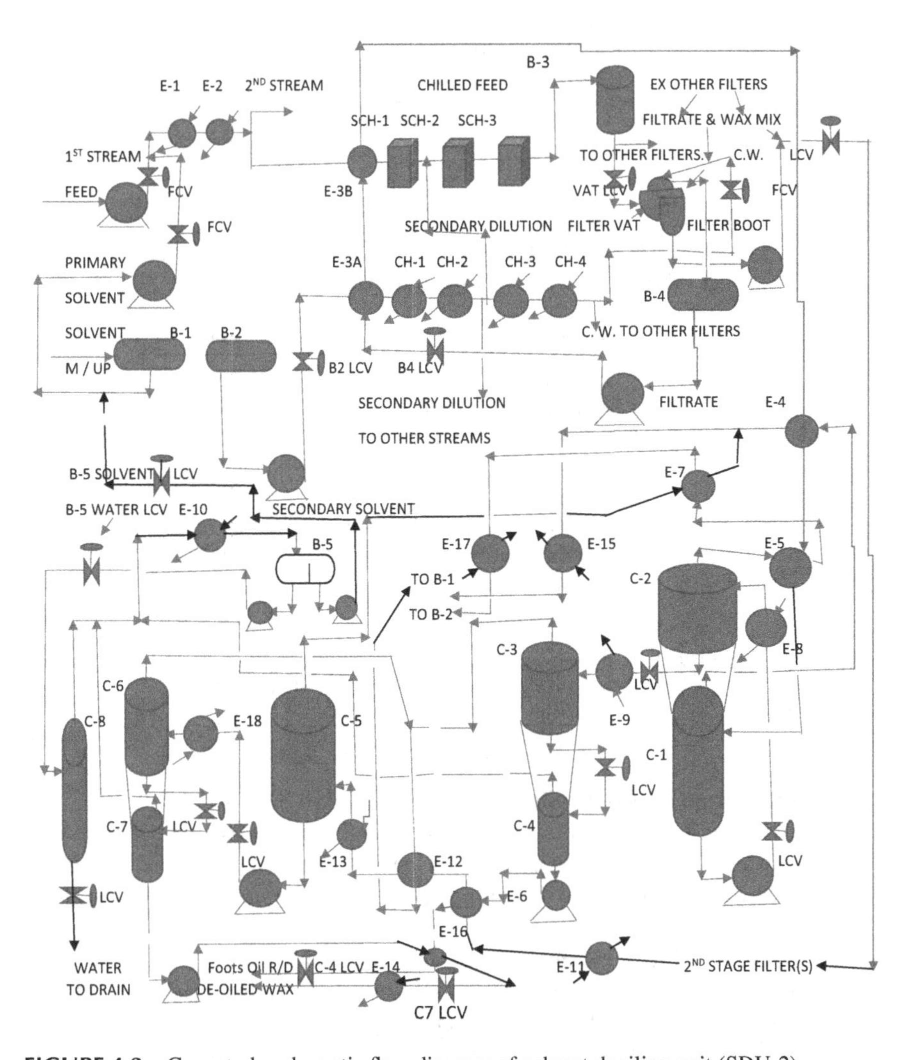

FIGURE 4.3 Case study schematic flow diagram of solvent deoiling unit (SDU-2).

TABLE 4.6
Case Study Operating Parameters in Solvent Deoiling Unit (SDU-2) for Paraffin Wax Production

Sl. No.	Parameter	Values
1	Primary Solvent Ratio, V/V (S:F)	1:1–1.5:1
2	Secondary Solvent Ratio, V/V (S:F)	1:1
3	Tertiary Solvent Ratio, V/V (S:F)	NIL
4	Dewaxing Filter Temp., Deg.C.	(+)3–5
5	Spray Solvent Ratio in Filter, V/V	1:1
6	Solvent Spray Temp. in Filter, Deg.C.	(–)3-(–)5
7	Filter Vat Level, %	30 Max.
8	Filter Vacuum, Bar	(–)0.85
9	Filter Blanketing Pr. Bar abs.	1.15–1.2

TABLE 4.7
Case Study Operating Parameters in Solvent Deoiling Unit (SDU-2) for MCW Production

Sl. No.	Parameter	Values
1	Primary Solvent Ratio, V/V (S:F)	2.0:1:2.5:1
2	Secondary Solvent Ratio, V/V (S:F)	1:1
3	Tertiary Solvent Ratio, V/V (S:F)	NIL
4	Dewaxing Filter Temp., Deg.C.	(+)3–5
5	Spray Solvent Ratio in Filter, V/V	1:1
6	Solvent Spray Temp. in Filter, Deg.C.	(–)3-(–)5
7	Filter Vat Level, %	30 Max.
8	Filter Vacuum, Bar	(–)0.85
9	Filter Blanketing Pr. Bar abs.	1.15–1.2

Note: As seen in Table 4.7, primary dilution ratio is higher due to feedstock viscosity is more than double in case of MCW production.

- **Typical properties of deoiled wax and foots oil in SDU-2:**
The desired qualities obtained in deoiled wax (for subsequent processing it in wax hydro-finishing unit to get best quality waxes as per specification) are shown in Table 4.8 along with expected qualities of by product, foots oil.
- **Case study utility consumptions in solvent deoiling unit (SDU-2):**
Solvent like MIBK is consumed due to losses from different leak sources, if any and major loss to the blanketing system of inert gas.

LP steam and MP steam are used in SDU; MP steam consumption becomes higher if no furnace is provided in foots oil-solvent circuit where solvent load is very high. Electrical power consumption is highest among all other

TABLE 4.8
Desired Properties of Deoiled Wax and Foots Oil

Sl. no.	Attribute	Properties	Paraffin Wax grade	MCW grade
1A	Deoiled Wax	- Oil content, %wt.	0.1–0.4	0.9–1.4
		- DMP, °C	55–58	70–85
		- Color, ASTM	1–1.5	1–2.5
		- Asphaltene content, PPM	<100	<300
1B	Foots Oil	- Pour Point, °C	15 to 21	18–24
		- Viscosity, cst @100°C	2.5–4.5	5–8

TABLE 4.9
Case Study Utilities Consumptions in SDU-2 for Paraffin Wax Production

Utilities per MT of Unit Feed Capacity	Consumption Rate
MIBK, Kg	1.2
LP (70 psi) steam, Kg	400
MP (200 psi) steam, Kg	600
Electricity, KWH	200
Circulating Water, M^3	60

TABLE 4.10
Case Study Utilities Consumptions in SDU-2 for MCW Production

Utilities per MT of Unit Feed Capacity	Consumption Rate
MIBK, Kg	1.5
LP (70 psi) steam, Kg	500
MP (200 psi) steam, Kg	800
Electricity, KWH	250
Circulating Water, M^3	70

process units discussed in Chapter 2. But in solvent deoiling unit, SDU-2 utility consumptions are lower with respect to solvent dewaxing unit, SDU due to the less severe process conditions as mentioned above. Typical average consumptions are given in Table 4.9 and 4.10 for paraffin wax and MCW production respectively in SDU-2.

- Metallurgy and investment cost:
  Though this unit also involves separation process without chemical reaction, the operating cost is very high as understood from the utilities' consumptions

figure from Table 4.9 and 4.10 because there are more numbers of sub-sections with high operating loads in each section. Accordingly, the number of pieces of equipment are also very high with some major capacity rotary equipment like compressors. But pressure of various sections remains in the range of two to five bar with exception in refrigeration section at about 15 bar, and temperature also ranges from ambient to about 180°C with exception in refrigeration/chiller circuits where temperatures remain sub-zero; thus, carbon steel pipe and fittings of normal grades with low carbon steel at sub-zero temperature are adequate for this unit, but equipment are large in numbers in this unit. This unit capacity generally is on lower side and thus unit investment cost is on the higher side. The approximate installed investment cost becomes 1.0 million USD per TMTA of unit feed capacity for a 100 TMTA feed capacity plant.

4.5.2 Solvent Deoiling of Aromatic Extracted Distillate (Raffinate)

If it were the oil with good lubricity as well as the wax quality also were good, then instead of carrying out direct solvent deoiling of the distillate, first solvent dewaxing with prior aromatic extraction (as per aromatic / asphaltene content in the distillate) is done to the oil in order to get the dewaxed oil to use it as lube base stock; the wax which is obtained as slack wax (called such due to having higher oil content in the wax at about 25%wt.) is taken as feedstock for processing in next unit, i.e. solvent deoiling unit (SDU-2) to get quality wax from there leaving the residual oil as foots oil which is very low in quantity as the main quantity of the oil has already been taken as lube base oil in the preceding unit, i.e. solvent dewaxing unit as mentioned above.

Also, when DAO is used as feedstock to get MCW grades of waxes, the flow processes, i.e. inclusion of aromatic extraction unit should be considered in the flow scheme even if there is no lube potential in the oil, in order to avoid aromatic compounds remaining in the product, deoiled wax.

The above scheme is represented through BFD, Figure 4.4.

- **Case study operating conditions in solvent deoiling using aromatic extracted distillate (Raffinate) as feedstock:**
  Here, as deoiling is done after dewaxing, all utilities consumptions and solvent consumption are on the higher side but against higher cost at these ends, value additions are also higher, i.e. two products, viz. DWO (as base stock after processing in subsequent hydro-finishing unit) as well as wax (paraffin wax or MCW after processing in subsequent hydro-finishing unit) can be obtained.

  The typical operating conditions in SDU in step (i) as shown above are similar as shown in Chapter 2 in Table 2.13 with 500N case for paraffin wax production and with 150BS case for MCW production, and in step (ii) as shown above, typical operating conditions in SDU-2 are similar as shown in Table 4.6 for paraffin wax production and Table 4.7 for MCW production.
- **Typical properties of deoiled wax and foots oil in SDU-2:**
  Due to repetition of similar operation, i.e. once for dewaxing in SDU and next for deoiling in SDU-2 as shown in step (i) and (ii) above, the combined utilities consumption would be higher but getting two premium products simultaneously

i. First step:

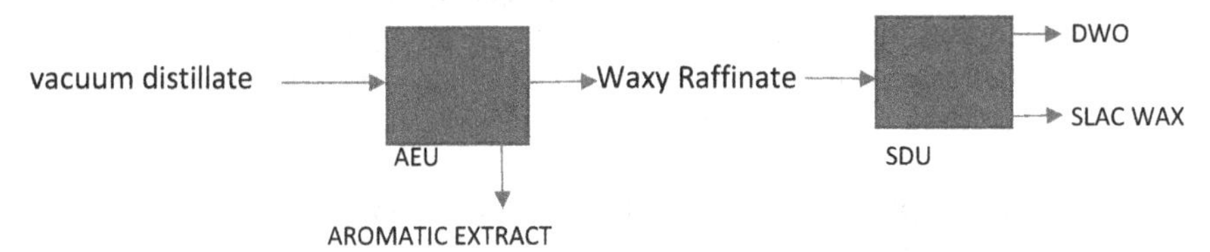

ii. Second step:

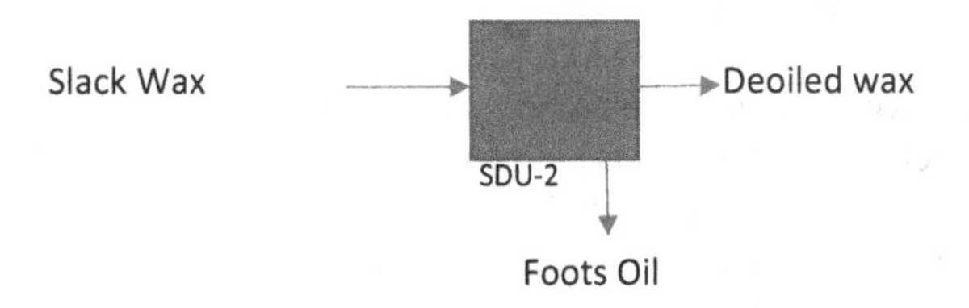

FIGURE 4.4 Solvent dewaxing and deoiling to produce both lube oil and wax.

offsets the enhanced cost on utilities. The combined utilities consumptions are shown in Table 4.11.

- **Utilities consumptions for the combined process of SDU and SDU-2:** In SDU, with aromatic extracted distillate as feed, dual solvent like MEK + toluene is used as explained in Chapter 2 while for the final wax production in SDU-2, only MIBK is used as solvent as mentioned above. As two products like DWO and wax are considered as output from this method, combined utilities consumptions since SDU plus SDU-2 should also be expressed per unit feed rate of SDU. Accordingly, the figures are shown in Table 4.11 as mentioned above.
- **Illustration:**
  i. Q. 100 MT of vacuum distillate is processed in AEU followed by SDU followed by SDU-2 as shown in Figure 4.4 above with raffinate yield

TABLE 4.11
Combined Utilities Consumptions in SDU and SDU-2

Utilities per MT of Unit Feed Capacity expressed in terms of SDU	Consumption Rate during paraffin wax production	Consumption rate during MCW production
MEK, Kg	1.5	1.6
Toluene, Kg.	1.5	1.6
MIBK, Kg	0.3	0.5
LP (70 psi) steam, Kg	350	360
MP (200 psi) steam, Kg	600	620
Electricity, KWH	200	210
Circulating Water, M^3	65	70

at 60%wt., DWO yield of 72%wt., slack wax oil content at 25%wt. To produce deoiled wax of paraffin wax quality, i.e. with oil content of about 0.3%wt., how much will be deoiled wax and by-products quantities?

Raffinate yield = 60 MT; hence, DWO yield = 60 × 0.72 MT = 43.2 MT.

i.e. Slack wax yield = 60 – 43.2 = 16.8 MT; its oil content = 16.8 × 0.25 MT = 4.2 MT.

Now, deoiled wax oil content = 4.2 × 0.003 MT = 0.0126 MT.

So, quantity of deoiled wax produced = 16.8 – 4.2 + 0.0126 MT = 12.6126 MT, say 12.6 MT.

Quantities of by-products: aromatic extract = 100 – 60 MT = 40 MT and foots oil = 4.2 – 0.0126 = 4.1874 MT, say 4.19 MT.

ii. Q. 24 MT/Hr. of waxy vacuum distillate is processed in SDU-2 as shown in Figure 4.1 earlier where 4 numbers of rotary vacuum filters are used as primary filters to produce wax with 5%wt. oil content with 25%wt. wax yield. How many secondary filters would be required for filtration of this primary wax for filtration to produce finished wax with oil content at 0.2%wt.?

A. Feed capacity of each primary filter = 24/4 = 6 MT/Hr.

Quantity of primary wax per each filter = 24 × 0.25/4= 1.5 MT/hr.

Oil removed from each primary filter = (24 – 6)/4 = 4.5 MT/hr.

Oil content in primary wax per filter = 1.5 × 0.05 = 0.075 MT/hr.

So, in the secondary filtration, processing rate is 1.5 MT/hr. with oil content to be removed is only marginally less than 0.075 MT/hr. per filter as the oil content in the final deoiled wax is only 0.2%wt.

Hence, requirement of secondary filter is only one no.

4.5.3 Solvent Deoiling of Oily Wax from Tank Bottom

Wax is lighter, i.e. its density is lower than the oil in equivalent boiling range; hence, wax in a waxy crude oil tank should float as a top layer inside the tank on long storage if it separates out from the oil; but it doesn't happen so because when, after long storage of waxy crude oil, the tank is emptied out, it is observed that tank bottom sludge contains a thick layer of wax including aromatics/asphaltene and other sludge material; this phenomenon indicates that aromatics/asphaltene concentrate, being heavier than major part of the oil with less concentration of aromatics in it, settles at the tank bottom during sludge formation and the naphthenic, poly-naphthenic and aromatic compounds with waxy side chains in them also being heavier than other waxes get miscible and absorbed in the bottom sludges.

The waxes in solution with main stream of oil in the tank are found to be of paraffin wax nature and the waxes in solution with the bottom sludge are to be of microcrystalline wax (MCW) nature.

So, when the tank bottom sludge waxes are to be processed to recover the wax in it, the processes as mentioned in sub-Section 4.5.2 earlier should be followed, i.e. first aromatic extraction should be done in AEU (aromatic extraction unit) with the tank bottom sludge as feed, then the raffinate produced should be either directly processed in SDU-2 or to be processed in SDU and the slack wax obtained in SDU is to be processed next in SDU-2. The operating conditions in AEU and SDU should be same as those shown in Chapter 2 respective sub-sections followed during bright stock, i.e. 150BS grade operations, and the operating conditions in SDU-2 should be followed as shown in Table 4.7. The product, MCW grade deoiled wax obtained from SDU-2 would be of qualities as shown in Table 4.8, but it is also observed that many a time, the drop melting point (DMP) comes in the lower range of DMP as in Table 4.8, i.e. at 70°C or marginally more or less thus after processing through wax hydro-finishing unit it leads to the production of finished MCW of type-A, B or C grade as shown in Table 4.3.

4.5.4 Solvent Deoiling of Hydrocracker Bottom

While two stage hydrocracker gives full conversion of feedstock leading to production of fuel oils only, the single stage hydrocracker without recycle generates residue at about 50%wt. of the feed. As the working principles of hydrocracker is to hydrogenate the aromatics while doing hydrocracking of the naphthene and paraffin, there is negligible presence of aromatics in the residue, i.e. residue contains a mix of naphthene, n-paraffin and iso-paraffin mostly. This composition of residue means it should be having lubricity, i.e. lube base oil can be recovered from this residue as well there would be presence of long-chain paraffins, i.e. waxy paraffin.

As lube potential generally becomes higher, i.e. at about 70 to 75%wt. than the remaining wax at about 30 to 25%wt., the residue is preferably processed in SDU to recover DWO which can be used lube base stock without processing in lube hydro-finishing unit as all sulfur and nitrogen are already removed in hydrocracking operation. The slack wax obtained from SDU can be further processed in SDU-2 to produce finished grade of paraffin wax of market specifications without processing in wax-hydro finishing unit due to reason as mentioned above.

Instead of producing both lube oil base stock and paraffin wax, some petroleum refineries may not like to produce wax and instead they want to convert the whole residue to LOBS as main product with fuel distillates as fete-accompli by-products and thus choose to process this hydrocracker residue in CIDW unit as detailed in Chapter 2, that too, use of HDT in CIDW unit can be skipped (as aromatics are already saturated in the hydrocracker) and instead processing in Iso-dewaxing and aro-sat (mild hydrogenation) sections of CIDW unit can be done.

4.5.5 Hydro Finishing of Deoiled Wax

The deoiled wax obtained from solvent deoiling process as mentioned in the sub-Sections 4.5.1, 4.5.2 and 4.5.3 above are dark in color, particularly when produced from heavier feedstock as mentioned in these sub sections.

Like DWO obtained from SDU as discussed in Chapter 2 needs hydro-finishing process to be followed to remove sulfur and nitrogen to improve color of the finished product, lube base oil as discussed in sub-Section 2.4.1.4 of Chapter 2, deoiled wax also should be treated in hydro-finishing unit to remove the nitrogenous and sulfur compounds from the deoiled wax. But, the oil content of deoiled wax is very negligible, then the question arises how there can be dark color in deoiled wax when the nitrogenous mainly and sulfur compounds party contribute to color of the product. Hence, contribution of these compounds to the color of wax is not that much high but the main reason of dark color of wax is due the presence of asphaltene particles which even in ppm concentration range make the color of the wax blackish/grayish.

It may be mentioned that these asphaltene particles are called refractory stock as these can't be hydrogenated by any conventional catalyst which are so far commercialized.

It is also found that the deoiled wax if passed through Ni-Mo catalyst (or alike), the asphaltene particles get adsorbed in the micro pores of the catalyst when the pores get wide opened up after raising the temperature of the catalyst; but in this way, the active surface area of the catalyst gets reduced resulting in availability of less area for hydrogenation and desulfurization reactions.

Hence, catalyst manufacturers have found it economical to offer a higher ratio of Ni-Mo catalyst as compared to feed processing rate to prolong the cycle length of operation with the same catalyst instead of offering alternative catalyst which were found to be impractical.

So, the principles of wax hydro-finishing as same as those for lube hydro-finishing as explained in sub-Section 2.4.1.4 of Chapter 2 except the catalyst to feed ratio is higher in wax hydro-finishing than in lube hydro-finishing. However, quantity of catalyst in wax hydro-finishing is always found to be much lower than in lube hydro-finishing, as conventionally, feed rate in wax hydro-finishing is found to be much lower than that in lube hydro-finishing as observed in all available commercial size plants.

- **Principles of operation of wax-HFU:**

  The process description for wax hydro-finishing is also same as in lube hydro-finishing except that in wax hydro-finishing feed enters from the bottom of the reactor and exits from the top of the reactor unlike reverse in lube hydro-finishing unit to take the advantage of better mixing of the feed and hydrogen with the catalyst while there is no chance of catalyst blowing up from the top of the reactor due to very high catalyst quantity, i.e. higher wt. of the catalyst whereas feed rate is very low, i.e. due to higher ratio of catalyst to feed as mentioned above.

  Also, the recycle gas sweetening circuit is not required in wax hydro-finishing unit due to very negligible oil content in deoiled wax resulting in very low sulfur content in the deoiled wax which can be taken care of by purging of H_2S time to time to fuel gas circuit from the sour recycle gas without investing in installation and operation of amine wash section for sweetening of sour recycle gas; rather, in wax hydro-finishing unit, continuous dosing of DMDS (di-methyl di-sulfide) to recycle gas is done by a dosing pump to maintain H_2S level of 50 to 100 ppm in the recycle gas.

  There is another difference of operation of wax hydro-finishing unit from lube hydro-finishing unit, i.e. unlike in the finished lube base oil, where the sulfur content in DWO was high and was not required to be removed completely as color contributor is mainly nitrogen, in wax hydro-finishing unit, sulfur content in deoiled wax is so low (due to negligible oil content in wax) that all sulfur get removed in the process and the finished wax becomes sulfur free which in turn results oxidation of the finished wax in the storage tank in long run thereby causing gum formation in the wax. To prevent this, an anti-oxidant, viz. DBPC (di-butyl-para-cresol) compound in about 10 to 30% strength in solution with molten wax is dosed in the finished wax rundown line (at the exit from the unit) @50 to 100 ppm.

Also, the storage temperature of wax should be kept at about 10°C higher than its DMP; hence, paraffin wax storage temperature should be at about 70°C to 80°C and MCW storage temperature should be about 90 to 95°C.

Note: from storage safety point of view, no product should be stored at 100°C to avoid boiling of water, if it ingresses by chance; if temperature becomes close to 100°C, then product rundown temperature from unit should be increased to about 105 to 110°C.

- **Case study process description of wax-HFU:**

With the above points in view, the process description of wax-HFU can be described as follows:

Deoiled wax is taken from the storage tank at a temperature of about 80°C to wax-HFU for paraffin wax grade and at about 85–95°C for MCW grade deoiled wax processing where the feed is pumped in two stages, viz. first by a low pressure pump of about 3 to 4 Kg/cm^{2}g discharge pressure which is then to heated through a heat exchanger by exchanging heat with finished rundown wax at a temperature of about 180°C to heat up the feed up to about 120–130°C which then passes through a series of feed filters to remove the dirt, if any, in the feed. Then clean feed then is pumped by a high pressure pump by a positive displacement pump (selected due to low flow but with high pressure) called feed booster pump to raise the discharge pressure up to about 120 Kg/cm^{2}g as necessary to reach the reactor (after passing through a series of heat exchangers with reactor effluent followed by further heating through an electric heater) from its bottom (as explained above) at a pressure of about 110 Kg/cm^{2}g to facilitate asphaltene particle adsorption into the catalyst pores over and above desulfurization and denitrification as discussed above; for this particular process of asphaltene adsorption along with conventional desulfurization and denitrification to happen, make-up hydrogen gas and recycle hydrogen gas also are required like in lube hydro-finishing unit for which on make-up gas compressor and one recycle gas compressor are used which discharge the respective gases to meet the feed at feed booster pump discharge at the downstream of feed filters as mentioned above and the feed-hydrogen gas mix heating to a temperature of about 260 to 280°C in case of paraffin wax deoiled wax processing and up to 300 to 340°C in case of MCW grade deoiled wax processing.

The reactor effluent gas-wax separation post reactor is similar to that in lube hydro-finishing, i.e. the reactor outlet effluent leaving from reactor top then enters the heat exchangers as mentioned earlier where it is cooled to about 120°C to 160°C depending upon grade of wax used. The cooled mixed stream then enters a flash separator vessel and unreacted hydrogen and some negligible quantity of light hydrocarbon vapors exit from the top of that vessel followed by further cooling in water cooler to send it low temperature flash vessel from where the gas comes out at a temperature of about 45°C which is then used as recycle gas as mentioned above without any amine wash treatment unlike in lube hydro-finishing unit (as discussed in Chapter 2) because sulfur content in deoiled wax is so low that the concentration of H_2S in the recycle gas does not become more than 100 ppm; as such due to minimum purging of this gas to fuel gas sometimes lead to concentration of H_2S in recycle gas below 50 ppm when rather DMDS dosing is to be started to maintain its minimum concentration in the recycle gas.

The bottom of the high temperature flash vessel as mentioned above is routed to a low-pressure flash vessel without further cooling. The bottom of the low-pressure flash vessel flows to a steam stripping column via a steam heater to heat it up to about 180°C and the bottom of the stripper moves by gravity to a vacuum stripper to remove the lighter hydrocarbon to take care of the flash point of finished wax. The vapor from both of the stripper and vacuum drying column is condensed in an overhead condenser followed by routing the condensed hydrocarbon to a vessel from where it is pumped to slop tank. The top vapor from high temperature low pressure flash vessel is routed to a water cooler followed by routing it to a vessel called low temperature low pressure flash vessel from top of which the gas is routed to fuel gas system of the refinery.

The bottom of the vacuum drying column is heat exchanged with feed as mentioned earlier or with stripper inlet feed as mentioned above depending upon heat exchanger circuit design in specific cases. The product stream after this heat exchanger is further cooled to its rundown temperature as mentioned above through a water cooler, either by cooling water or by warm water, and pumped to finished LOBS storage tank.

The case study schematic flow diagram for the process depicting stream flows, process control and heat exchangers network is shown in Figure 4.5.

- **Case study operating conditions of HFU:**

Operating parameters to control the main reaction process are given in Table 4.12. Same as

- **Desired product qualities and yield:**
  To meet the product specifications as shown in Table 4.1 and 4.3 are the ultimate objectives to be achieved after processing deoiled wax in wax-HFU to get the finished wax.

  As the catalytic reactions in wax-HFU are mild ones, yield of the product is nearly same as the feed rate, viz. yield of about 99%wt. is achieved in wax-HFU but due to negligible undesired cracking of the deoiled wax into oil, the final oil content of finished wax increases by 0.1 to 0.2%wt. due to which oil content of deoiled wax is kept lower by 0.1 to 0.2%wt. from the specification.

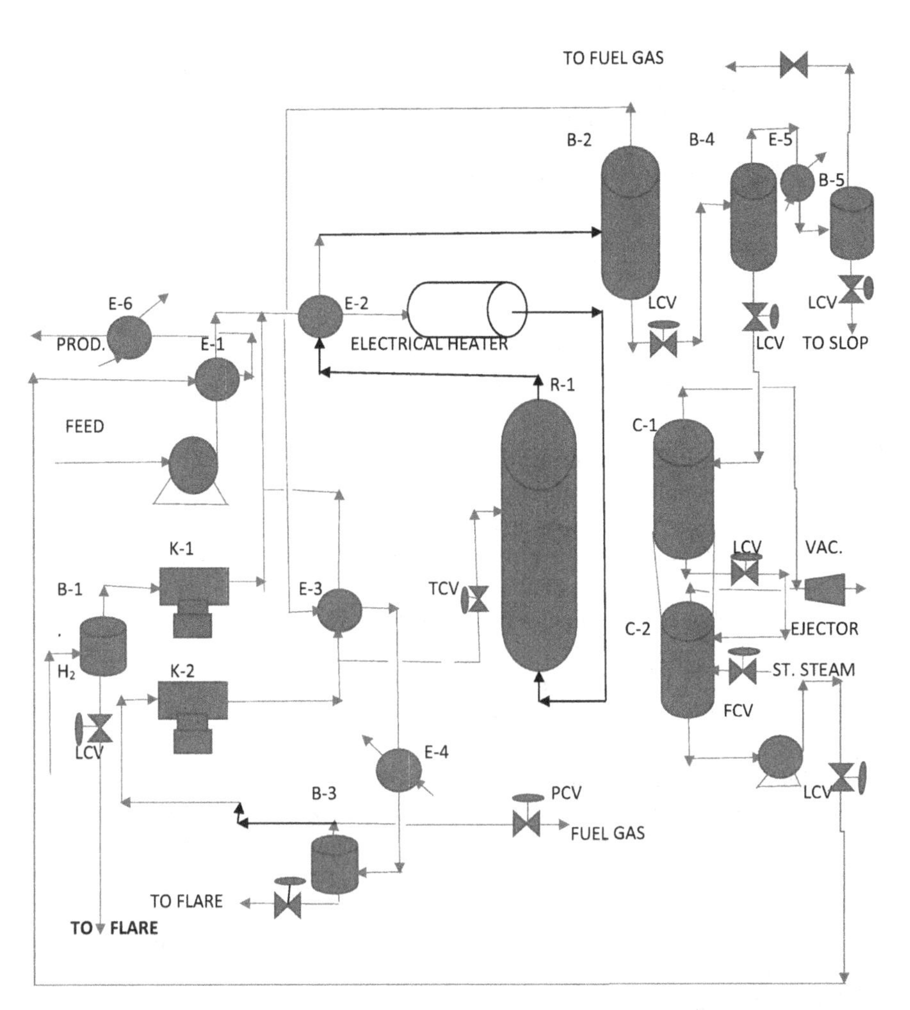

FIGURE 4.5 Case study schematic flow diagram of wax-HFU.

TABLE 4.12
Typical Operating Parameters of HFU

Sl. No.	Parameter	Paraffin Wax	MCW
1	Make-Up H_2 Gas to Feed ratio, Nm^3/m^3	5:1	10:1
2	Recycle Gas to Feed ratio, Nm^3/m^3	50:1	75:1
3	Furnace outlet/Reactor Temperature, °C	260	300
4	Reactor Pressure, Bar	110	110
5	Reaction space velocity, Hr^{-1}	0.5	0.5
6.	Make-up H_2 purity, %wt.	75 min.	75 min.
7.	Recycle H_2 purity, %wt.	75 min.	75 min.

TABLE 4.13
Utility Consumptions in Wax-HFU

Utilities per MT of Unit Feed Capacity	Consumption Rate
Hydrogen, Kg	1.3
LP (70 psi) steam, Kg	20
MP (200 psi) steam, Kg	40
Fuel, Kg	0
Electricity, KWH	30
Circulating Water, M^3	10

- Case **study utilities consumptions in Wax-HFU:**
It is reaction unit unlike physical separation units as described earlier above. In the reaction hydrogen is required as it is a hydrotreating unit, but consumption of hydrogen is low as it is a mild hydrotreating unit, however requiring high pressure reaction section of about 110 bar.
The utility consumptions are given Table 4.13.

As shown in Table 4.13, hydrogen consumption rate is higher due to more purging out of sour gas instead of providing gas sweetening circuit.

- **Metallurgy and Investment cost of Wax-HFU:**
As mentioned above, the unit being a reaction unit requires a high-pressure reactor of about 110 bar pressure. The unit contains two sub-sections with two sub-sections at high pressure of about 110 bar, i.e. first one is reaction section including reactor, two hydrogen compressors and the second section is low pressure section of about 5 bar pressure up to a vacuum column for product drying associated with piping, heat exchangers and pumps. Due to having reaction sections at high pressure of about 110 bar and at a temperature of 260 to 300°C, alloy steel (austenitic stainless steel) in high pressure reactor, heat exchangers, furnaces are required to be used. However, the unit is a compact unit with minimum equipment unlike physical separation processes as described earlier above.
With the above, the approximate installed investment cost becomes two million USD per TMTA of unit feed capacity for 50 TMTA plant capacity.
- **Illustrations:**
  i. Q. Wax-HFU is designed to process paraffin wax feed @5 MT/He. with asphaltene content of 100 ppm through a bed of 9 MT Ni-Mo catalyst which provides a catalyst cycle length of two years with 8000 hrs. of operation per year. If instead of paraffin-wax grade feed, MCW grade of feed is processed @5MT/hr. with asphaltene content of 300 ppm in the feed, then what would be cycle length of the catalyst?
  A. Total asphaltene content (Q) in the paraffin-wax grade which would be adsorbed in the catalyst bed is given by:

$$Q = 5000 \times 100 \times 10^{-6} \times 8000 \times 2 = 8000 \text{ kg.} = 8 \text{ MT.}$$

So, in MCW grade wax feed processing, maximum 8 MT of asphaltene adsorption can be allowed. Hence, the cycle length (L) of the catalyst in this case would be given by:

$L = 8000/ (5000 \times 300 \times 10^{-6}) = 5333.333$ hrs., i.e. 8 months only.

ii. Q. Capital investment cost for a 50 TMTA capacity of paraffin wax plant is 100 million USD. If the capacity increases to 70 TMTA, what would be the capital investment cost?

A. Thumb rule formula for evaluating of capital investment cost with varying capacity is given by:

$$P_2/P_1 = (C_2/C_1)^{0.6}, \text{ i.e. } P_2 = P_1 \times (C_2/C_1)^{0.6}.$$

Where, P_2 and P_1 are capital investment costs at plant capacity of C_2 and C_1 respectively.

Hence, $P_2 = 100 \times (70/50)^{0.6}$
$= 100 \times 1.224$ million USD.
$= 122.4$ million USD.

So, new capital investment cost is 122.4 million USD.

Note: though this illustration is given in this section, it is applicable to all sections as mentioned against each process unit.

4.6 WAX DERIVATIVES AND APPLICATION

Wax derivatives can be obtained in three ways as below.

- By physical blending of waxes of different grades:
  To meet specific market requirement, waxes of different grades are often blended, for example, paraffin wax and MCW can be blended to obtain desired characteristics as required for the demand product.
  Sometimes synthetic waxes like PE wax, butyl rubber, poly-iso-butylene, ethylene vinyl acetate co-polymers, resins with or without anti-oxidant as additive are blended with natural waxes to produce the finished wax as end product to meet market requirement.
- Fractional distillation of wax of finished grade(s):
  Wax can be fractionated to get a good number of fractionated products to use these for specific purposes.
- By chemical reactions of waxes of finished or un-finished grades:
  Slack wax or finished wax either of paraffin wax grade or MCW grade can be processed in iso-dewaxing section of CIDW unit as mentioned in Chapter 2 to isomerize the n-paraffin chains in the wax to reduce the pour point to the level of meeting the point of lube base oil. Instead of using costly finished grades of wax, it is more economical to use cheaper slack wax as feed as oil content in the slack wax is not more than 20–25%wt.
  Installing the HDT section at upstream of iso-dewaxing section depends on the quantity of oil present in the slack wax. Using aro-sat section at downstream of iso-dewaxing section also depends upon how much aromatics are found at the product of iso-dewaxing section.

While making lube base oil from wax, it may be noted that some by-products of fuel properties would be generated as explained in CIDW unit in Chapter 2.

Wax can also be used to produce very valuable product like nano-carbon for use in the industries.

5 Energy Optimization in Dewaxing and Deoiling Process

Broadly there are four energy optimization techniques commercialized so far for dewaxing and deoiling process as follows:

a) Application of pinch technology and heat integration.
b) Simultaneous dewaxing and deoiling technology.
c) Membrane separation.
d) Catalytic process (developed only for dewaxing process).

Both dewaxing and deoiling processes used for production of lube base oil and wax respectively are based on physical separation process (except the catalytic dewaxing process for dewaxing) and these are energy intensive due to a large amount of solvent is required to carry out the separation of wax from oil and vice versa and due to, again, this solvent being recovered by distillation to recirculate the solvent to the feed to continue the separation process which consumes a lot of heat energy to ensure distillation and hydraulic energy to ensure complete flow of the oil (or wax) solvent mixture through a large equipment network for both the circuits, viz. oil-solvent circuit and wax-solvent circuit. These two circuits are required both for lube base oil and wax production as discussed in Chapter 2 and 4 respectively. Now-a-days catalytic dewaxing process has come into play which is energy economic as compared to solvent dewaxing as discussed in Chapter 2, but for production of wax from petroleum sources, there is no commercial route for catalytic process and only solvent extraction process is followed for wax production as discussed in Chapter 4.

Hence, improvement in the solvent extraction separation process for recovery of wax from the waxy distillate is the present solution to reduce energy consumption for wax production. Pinch technology is used for this purpose in all heat exchangers network, and its application in wax production process is most important for energy savings point of view.

While using pinch technology, there is a mathematical approach of heat integration between two process units' streams with due consideration to HAZOP (hazard and operability studies) application from safety point of view. The heat integration concept among all process streams inside a unit and/or considering streams between two process units adjacently located is much encouraging approach for best application of pinch technology.

DOI: 10.1201/9781003334422-5

Many refineries produce lube oil as well as wax also through separation process as mentioned above. In such situation, the slack wax produced from solvent dewaxing unit enters the solvent deoiling unit which again consume solvent to dissolve the slack wax in it to recover the quality finished wax by evaporating the solvent out of it after residual oil separation from the slack wax; this method causes duplicate consumption of solvent making it energy intensive process. To avoid such situation, if the two process units are integrated into a single process unit where the slack wax-solvent mixture obtained from the filters in dewaxing operation (to produce lube oil) enters directly the deoiling filter(s) with or without addition of further solvent, then there would be lot of energy savings in that scenario. This process is called simultaneous dewaxing-deoiling process and finds commercial application as being followed by some refineries.

In addition to above, there are another approach foe energy optimized operation, i.e. membrane separation process which facilitates less consumption of solvent due to prudent use of membrane.

All the above four technologies are discussed as follows.

a) Application of pinch technology and heat integration:

Before discussing the case studies of application, first let us recapitulate the concept of pinch technology as follows:

- Pinch point is a temperature.
- It divides the temperature range into two regions.
- Heating utility can be used only above the pinch and cooling utility only below the pinch.
- A heat exchanger network obtained using the pinch design is a network where no heat transfer occurs from a hot stream whose temperature is above the pinch to a cold stream whose temperature is below the pinch.

Let us explain through following an example where there are four heat exchangers with one hot stream enters a heat exchanger at a temperature denoted by $T_{H1, in}$ and leaves it at a temperature denoted by $T_{H1, out}$ after heat exchanging an amount of Q_1 with the cold stream. Next another hot stream enters a second heat exchanger at a temperature of $T_{H2, in}$ and leaves it at a temperature of $T_{H2, out}$ after heat exchanging an amount of Q_2 with the same cold stream.

The above can be shown diagrammatically in Figure 5.1.

If the point B and C in Figure 5.1 are joined, the composite line obtained by joining A, B, C and D is called composite curve. This composite line is called cooling line where the line of cold stream is called heating line.

Similarly, there can be composite curve for the heating line if there are more than one cold stream. Now, if such composite curve is drawn for the cold streams also, and next if that cold composite curve is pushed up and then to right, it would result in increased heating and cooling by the same amount as pushed; it would also result in getting the smallest delta T region between hot and cold composite curves, and this smallest deltaT is called pinch point.

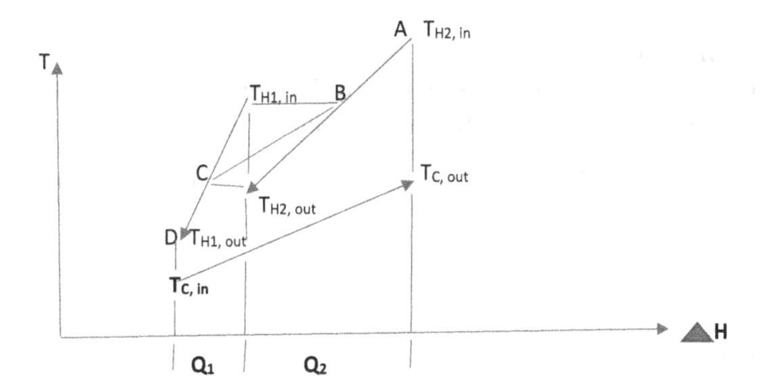

FIGURE 5.1 Pinch point diagram in heat exchangers.

Legend: T is the temperature and ▲H is the enthalpy lines, and Q_1 and Q_2 are heat exchanged by the hot stream one and hot stream two respectively with the cold stream.

TABLE 5.1A
Case Study Utilities Consumption of Deoiling Unit for Paraffin Wax Production without Application of Pinch Technology

	Consumption Rate	
Utilities per MT of Unit Feed Capacity	**Refinery 1**	**Refinery 2**
MIBK, Kg	1.3	1.3
Fuel oil, Kg/MT	40	0
LP (70 psi) steam, Kg	400	500
MP (200 psi) steam, Kg	500	700
Electricity, KWH	200	200
Circulating Water, M^3	70	70

In practice, there should be a reasonable low value of this pinch point as otherwise the area below the cold composite curve would increase which means more heat exchange surface area leading an un-economic situation.

By application of pinch technology as mentioned above, while hot utility, viz. steam consumption (or fuel oil/gas consumption where furnace is used) can be reduced through optimized heat integration, flow of circulating cooling water also can be reduced as the cold stream(s) come out of the heat exchanger network with less temperature.

- Benefit on commercial application of pinch technology along with heat integration:

In India, pinch technology application found to be started since 1992 in solvent dewaxing unit of a refinery which resulted in steam energy savings by about 15%, energy consumption data were collected for the deoiling units of the refineries in India where two units commissioned before 2002 without application of pinch technology and the third unit was commissioned in 2015 with the application of pinch technology. The case study data are shown in Table 5.1A and 5.1B:.

TABLE 5.1B
Case Study Utilities Consumption of Deoiling Unit for Paraffin Wax Production with Application of Pinch Technology

Utilities per MT of Unit Feed Capacity	Consumption Rate
MIBK, Kg	1.2
LP (70 psi) steam, Kg	400
MP (200 psi) steam, Kg	600
Electricity, KWH	200
Circulating Water, M^3	60

b) Simultaneous dewaxing and deoiling-a case study:

Some refineries have both dewaxing unit for lube base oil production as well as deoiling unit where the by-product of dewaxing unit, i.e. slack wax is used as feedstock to another unit, i.e. deoiling unit to produce quality on-spec wax of either paraffin wax grade or micro-crystalline wax (MCW) grade.

It is observed that in the dewaxing process, from the dewaxing filters two streams come out, one dewaxed oil-solvent mix and the other is slack wax-solvent mix; these two streams subsequently pass through respective solvent recovery circuits from where solvent free dewaxed oil (also called lube base oil when qualities meet spec.) is obtained and solvent free slack wax is obtained. This slack wax contains residual oil which prevents it to sell as finished quality and thus sold as secondary fuel to the customers or sent to further energy consuming deoiling unit where it is processed like dewaxing unit except with different set of operating parameters with the objective to get wax as the main product and oil here becomes a by-product, called foots oil (sold as secondary fuel to other process units) which has quality mis-match with respect to dewaxed oil as mentioned above. So, it is evident, while making finished wax, the production procedure involves energy consumption to recover solvent free slack wax in dewaxing unit and again this solvent free slack wax is added with further solvent in deoiling unit where deoiled wax-solvent obtained from the filters is again sent solvent recovery circuit to produce solvent deoiled wax, i.e. quality wax.

To avoid duplication of energy consumption as mentioned above, such refinery should make their dewaxing unit and deoiling unit integrated where slack wax-solvent mix obtained from the dewaxing filters is sent directly to deoiling filters with or without addition of further solvent; sometimes, this slack wax-solvent is sometimes passed through another set of chillers after injection of additional solvent to it before sending the same to deoiling filters. This process is called simultaneous dewaxing and deoiling which saves lot of energy for the refinery which has both lube base oil as well as wax also as main products. The energy saving is further improvised with application of pinch technology (as mentioned earlier) in this integrated process. Refineries of today follow this principle of simultaneous dewaxing and deoiling only; even the refineries, having only dewaxing unit to produce lube base oil, install a deoiling section in this dewaxing unit to carry out the above-mentioned simultaneous

TABLE 5.2A
Case Study Operating Parameters in Dewaxing and Deoiling Sections

Sl. No.	Parameters	Dewaxing	Deoiling
1	Feed Type	Distillate or Raffinate	Slack wax-solvent mix
2	Feed KV, cst@100°C	10–11 (distillate); 8–8.5 (Raffinate)	Equivalent slack wax KV of 4.5–5.5 as obtained
3	Solvent ratio (v/v)		
	-primary	1:1	0.5:1
	Secondary	1:1	NIL
4	Chillers outlet temp.		
	Profile, °C	20/10/(−)15	20/10/(−)3 to (+)3
5	Primary Filter wash solvent ratio (v/v)	1:1	1:1
6	Secondary filter temp., °C	Not applicable	(+)3 to (+)6
7	Secondary filter solvent ratio, v/v	Not applicable	1:1

TABLE 5.2B
Case Study Operating Parameters in Dewaxing and Deoiling Sections

Sl. No.	Parameters	Dewaxing	Deoiling
1	Feed Type	Distillate or Raffinate	Slack wax-solvent mix
2	Feed KV, cst@100°C	34–36 (distillate); 29–31 (Raffinate)	Equivalent slack wax KV of 6.5–7.5 as obtained
3	Solvent ratio (v/v)		
	-primary	3.5:1	1:1
	Secondary	NIL	NIL
4	Chillers outlet temp.	20/10/(-)15	20/10/(+)3 to (+)5
	Profile, °C		
5	Primary Filter wash solvent ratio (v/v)	1:1	1:1
6	Secondary filter temp., °C	Not applicable	0 to (+)3 to (+)5
7	Secondary filter wash solvent ratio, v/v	Not applicable	1:1

Note: The wax-solvent mix from the primary filter in deoiling section directly enters the secondary filter(s) without addition of any further solvent, i.e. for just re-filtration of wax-solvent mix obtained from the primary filters of the dewaxing section; only wash solvent is sprayed in the secondary filter(s).

dewaxing and deoiling operations to produce quality wax instead of selling slack wax at a very low price or using it as a secondary fuel in their other process units replacing more cheaper fuel oil.

- **Case study operating conditions:**
  - Operating parameters for Production of 500N base oil and paraffin Wax and the same for production of 150BS base oil and micro-crystalline wax (MCW) are shown in Table 5.2A and 5.2B respectively.

- Energy savings in simultaneous dewaxing and deoiling:

It is seen from Table 4.6 in Chapter 4 that dilution solvent ratio was 1.5 to 1:1 in the dedicated deoiling unit for paraffin wax production whereas in the integrated unit, solvent ratio is only 0.5:1 as seen in the Table 5.2A; similarly, it is seen from Table 4.7 in Chapter 4 that dilution solvent ratio was 2 to 2.5:1 in the dedicated deoiling unit for MCW production whereas in the integrated unit, the dilution solvent ratio is only 1:1 as seen from Table 5.2B. As far as filter wash solvent ratio is concerned though the ratios are same in both dedicated deoiling unit and integrated deoiling unit, the total number of filters in integrated deoiling unit are generally less by one or two as compared to that in dedicated deoiling unit due to total less solvent load in the integrated unit which results in decreased wash solvent load in the integrated deoiling unit.

The resulting decreased energy consumption in integrated deoiling unit as reflected by lower utilities consumptions has been compared to those in dedicated deoiling unit and the case study results are reproduced Table 5.3A and 5.3B for paraffin wax and MCW respectively.

c) **Membrane separation:**
- Objectives of the membrane separation section:

TABLE 5.3A
Case Study Comparison of Energy Consumption per MT of Feed in Case of Paraffin Wax Production

Sl. No.	Parameters	Dedicated deoiling	Integrated deoiling
1	Solvent used	MEK and Toluene-1:1	MIBK
2	Solvent consumption, kg.	1.4	1.2
3	LP (70 psi) steam, kg.	600	400
4	MP (200 psi) steam, kg.	800	600
5	Electricity, KWh.	220	200
6	Cooling water, M^3.	80	60

TABLE 5.3B
Case Study Comparison of Energy Consumption per MT of Feed in Case of MCW Production

Sl. No.	Parameters	Dedicated deoiling	Integrated deoiling
1	Solvent used	MEK and Toluene-1:1	MIBK
2	Solvent consumption, kg.	1.5	1.3
3	LP (70 psi) steam, kg.	600	450
4	MP (200 psi) steam, kg.	800	650
5	Electricity, KWh.	220	200
6	Cooling water, M^3.	80	60

Solvent dewaxing units have different operating constraints like refrigeration system loads to get the desired filter feed temperature, solvent recovery distillation loads and hydraulic limitation leading to huge hardware system to achieve desired production capacity. Degrees of refrigeration and solvent ratios decides the number of filters to support the production rate. Lastly, the dewaxed oil recovery section which handles the largest volume of solvent, may be constrained by the heat input possible through the steam heaters or furnaces.

The use of organic membranes to selectively recover solvents from the dewaxing process can significantly increase base oil production by simultaneously debottlenecking the refrigeration, filtration and solvent recovery sections of a dewaxing and deoiling unit. The process uses a proprietary polyimide membrane to remove the solvent, methyl-ethyl ketone (MEK) and toluene or methyl-iso-butyl ketone (MIBK) from the dewaxed oil or deoiled wax, as the case may be, leaving the filtrate receiver vessel. The solvent is recovered at or near the dewaxing/deoiling temperature and can be recycled to the dewaxing/deoiling process without the need of additional cooling thereby saving refrigeration energy. This additional solvent decreases the filter feed viscosity, thereby increasing the filtration rate ultimately leading to increased production rate. MOBIL and W. R. Grace, a world leader in membrane separations, jointly developed this membrane separation technology. The first commercial membrane unit started up successfully in May 1998 in MOBIL's Beamount Texas refinery. Base oil production from application of this technology increased by about 20% as a result of good selection of the membrane and ancillary equipment installed. The capital cost of the membrane unit was about one-third of that which otherwise would have been required for a conventional expansion/debottlenecking of the same magnitude.

- Membrane separation principles and techniques:

  A membrane is a selective barrier that permits the separation of certain constituents of a mixture of gas or liquid by combination of sieving and sorption diffusion mechanism. Membrane has gained an important place in chemical technology and have a broad range of applications. Membrane processes separate molecules on the basis of size and molecular weight. It's a pressure driven process. There are different types of membrane separation techniques, viz. reverse osmosis, ultrafiltration, electrodialysis, nanofiltration, micro-filtration and pervaporation. Pervaporation and reverse osmosis processes are generally applied in the industries for oil-solvent separation. Pervaporation is membrane separation process, involving the partial vaporization of liquid mixture through a dense membrane for selective permeation of one or more components from a liquid mixture, whose downstream side is usually kept under vacuum while upstream may be kept under high pressure depending upon the liquid mixture constituents' properties and type of membrane selected, but in pervaporation main driving force is activity difference (though the driving force may be pressure differential in addition to activity difference towards the membrane absorption) while in reverse osmosis, the driving force is a differential pressure of 10 to 100 Bar. A simple sketch of membrane separation is shown in Figure 5.2.

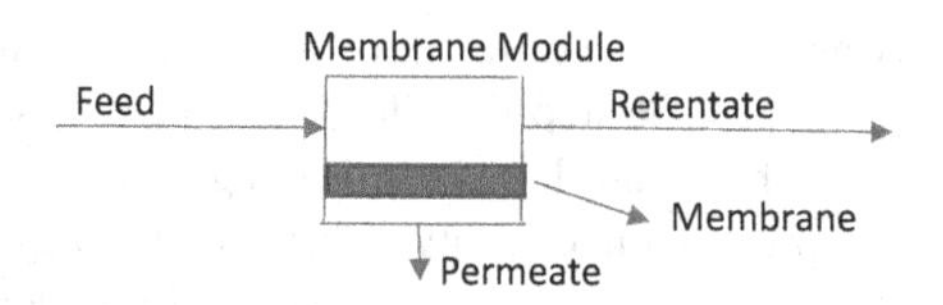

FIGURE 5.2 Sketch of a membrane module.

- **Theoretical principles for permeation through membrane:**
  The starting point for the mathematical expression of diffusion through membranes is the proposition, based on thermodynamics, that the driving forces of pressure, temperature, concentration and electrical potential are interrelated and the overall driving force producing movement of a permeate is the gradient in its chemical potential. Thus, the flux, J_i ($g/cm^2.s$), of a component I, is described by the simple equation as below:

$$J_i = (-)\, L_i\, d\mu_i/dx \qquad (5.1)^{[18]}$$

Where, $d\mu_i/dx$ = chemical potential gradient of component I and L_i = coefficient of proportionality (not necessarily constant) linking this chemical potential driving force to a flux.

This approach is extremely useful as many processes involve more than one driving force, for example, both pressure and concentration in reverse osmosis.

Chemical potential of a species in solution is expressed as,

$$\mu_i = \mu_i^0 + RT\ln a_i = \mu_i^0 + RT\ln(\beta_i\, x_i) \qquad (5.2)^{[18]}$$

Where x_i = mole fraction of component I in the solution, β_i = activity co-efficient of component i and μ_i = chemical potential of component i.

- Membrane process description:

Membrane process description is shown schematically in Figure 5.3.

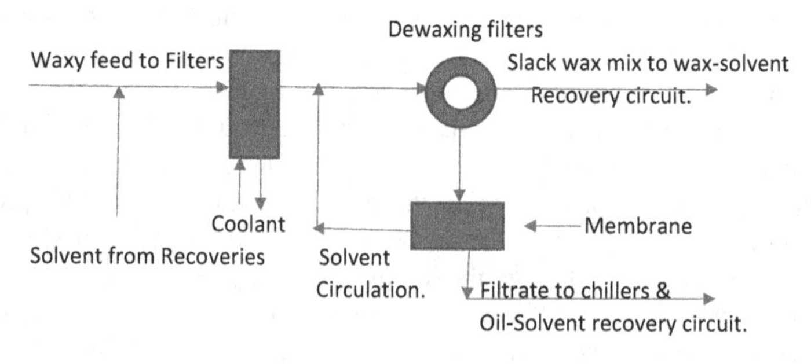

FIGURE 5.3 Solvent dewaxing with membranes.

As seen in the Figure 5.3, the dewaxing unit feed is first cooled to the target temperature, then dewaxed on rotary drum filters in the manner as followed in a dewaxing/deoiling unit explained in Chapter 2 and 4. The MEK/toluene/lube oil filtrate or MIBK/foots oil filtrate from the rotary drum filters of the dewaxing or deoiling unit, as the case may be, is fed to the membrane unit (via the filtrate receiver vessel) at a pressure between 30 to 45 bar. Membrane system temperature depends upon the upstream filtration temperature and typically ranges from (−)18 to (+)3°C. The membrane unit separates the filtrate from the solvent; the retentate stream after membrane separation is the dewaxed oil or foots oil, as the case may be, while the recovered solvent is called a permeate. The permeate exits the membrane at essentially atmospheric pressure, and is pumped directly to the dewaxing/deoiling chilling train as dilution solvent. The retentate exits the membrane unit and is let down to the conventional solvent recovery section where solvent load is very less due to prior separation of bulk solvent in the membrane unit, thus saving a lot of energy in the solvent recovery circuit due to very less consumption of heating system in the circuit.

A schematic diagram of a membrane separation unit with multiple membrane modules is shown in Figure 5.4.

As seen in Figure 5.4, the dewaxed oil/foots oil leaving the filtrate receiver is first sent to a bank of conventional cartridge type filters to remove dirt scale and any wax crystals which may be present in the feed stream. The feed is then pumped to a set of membrane housings, all operating in parallel, which contain the spiral wound membrane modules themselves. Each membrane housing contains multiple membrane modules which are connected in series by coupling their respective tubes. The permeate collected by the modules is pressured to a permeate receiver and then is pumped to the dewaxing/deoiling chilling train. The retentate, which is essentially at the membrane charge pump discharge pressure due to low axial pressure drop in the

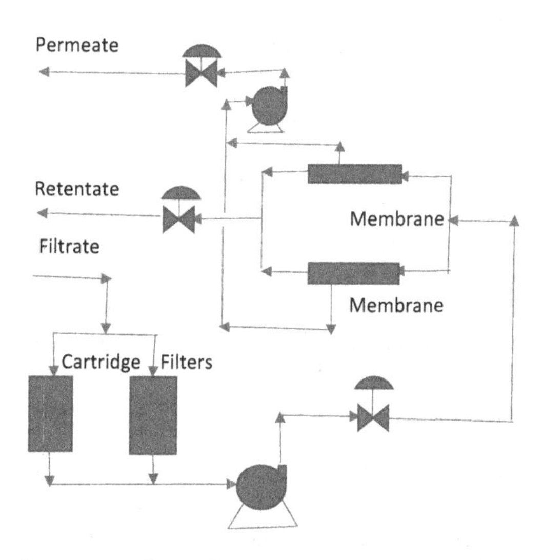

FIGURE 5.4 Membrane separation unit.

modules, is led down to the double pipe heat exchangers to exchange heat with feed stream of dewaxing/deoiling section.

The design of the membrane unit and, in turn, the amount of solvent recovery by the membranes, is optimized with respect to the individual limitations of the plant. All of the equipment in the membrane section is conventional with the exception of the membrane modules themselves.

Regarding the spiral wound modules, it may be mentioned that within the modules, the feed flows under pressure through a carefully designed feed channel where some of the solvent selectively permeates the membranes while the oil with remaining less solvent are rejected. The solvent passes through the semi permeable membranes at a rate controlled by the difference between the feed pressure and osmotic pressure of the system. Lube oil or foots oil, as the case may be, will also transport across the membrane under diffusion control at a relatively slow rate dictated by the inherent membrane selectivity. The clean permeate, which is at essentially atmospheric pressure, flows radially through the membrane envelope and is collected in the central support tube of the module. External to the module, it flows to the permeate surge vessel which is then pumped back to the dewaxing unit/deoiling unit for subsequent use as cold dilution solvent. The large radial pressure drop is supported entirely by the membrane and its associated backing material, which have a combined thickness of about 100–350 microns. The permeate product has an oil content of 0.1 to 0.6%wt. depending upon the viscosity grade of dewaxed oil or foots oil (equivalent to viscosity grade of wax) in the filtrate. The retentate rate and composition is dependent upon the operating severity of the membranes.

High solvent selectivity can be maintained indefinitely when the membrane modules are operated within a well-defined temperature window. The membrane has excellent solvent resistance and acceptable thermal stability. The spiral wound membrane modules are structurally robust and have a projected life more than two years. The feed pressure of 30–45 bar provides an acceptable balance between permeate production rate and membrane aging. The higher pressure can be used to increase permeate production at the expense of decreased membrane service life.

d) Catalytic process (for dewaxing):

The utilities consumption for catalytic dewaxing as in the Table 2.20 of Chapter 2 has been reproduced as Table 5.6 and compared with combination of utilities consumption in solvent dewaxing and hydro-finishing as in Table 2.15 and Table 2.17 of Chapter 2 (reproduced as Table 5.4A and 5.4B respectively), because utilities consumption are the reflection of energy consumption in the processes.

Combined utilities consumption of Table 5.4A and B in Chapter 2 represents the total utilities to be consumed for production of finished base oil from lube raffinate as computed; the utilities consumption through conventional process is shown in Table 5.5.

Now, if we compare Table 5.5 and 5.6, it is seen that all the common utilities consumption as in Table 5.5 are much higher than in Table 5.6 except hydrogen consumption which is higher in case of Table 5.6; also, in case of Table 5.6, there is fuel

TABLE 5.4A
Case Study Utilities Consumption in Solvent Dewaxing

Utilities per MT of Unit Feed Capacity	Consumption Rate
MEK, Kg	1.5
Toluene, Kg	1.5
LP (70 psi) steam, Kg	250
MP (200 psi) steam, Kg	400
Electricity, KWH	130
Circulating Water, M^3	70

TABLE 5.4B
Case Study Utilities Consumption in Lube Hydro-finishing (HFU)

Utilities per MT of Unit Feed Capacity	Consumption Rate
Hydrogen, Kg	1
LP (70 psi) steam, Kg	25
MP (200 psi) steam, Kg	45
Fuel, Kg	10
Electricity, KWH	40
Circulating Water, M^3	16

TABLE 5.5
Combined Utilities Consumption through Conventional Processes

Utilities per MT of Unit Feed Capacity	Consumption Rate
MEK, Kg	1.5
Toluene, Kg	1.5
LP (70 psi) steam, Kg	245
MP (200 psi) steam, Kg	435
Electricity, KWH	160
Circulating Water, M^3	80
Hydrogen, Kg	0.9

consumption of 25 kg/MT which is not seen in Table 5.5. But, 25 kg of fuel is equivalent to 340 kg of MP steam; so even if we add this amount of MP steam to the value of MP steam in Table 5.6, we find that MP steam consumption in case of Table 5.5 higher than in Table 5.6.

TABLE 5.6
Utilities Consumption in Catalytic Dewaxing Unit

Utilities per MT of Unit Feed Capacity	Consumption Rate
Hydrogen, Kg	7.5
LP (70 psi) steam, Kg	40
MP (200 psi) steam, Kg	50
Fuel, Kg	25
Electricity, KWH	50
Circulating Water, M^3	25

Note: Combined utilities in Table 5.5 have been computed based on 72%wt yield in solvent dewaxing unit.

Now we find that there are MEK and toluene consumption of 1.5 kg each in case of Table 5.5 which is not there in case of Table 5.6 but on the other side, there is additional hydrogen consumption of 6.6 kg in case of Table 5.6 as compared to Table 5.5. the financial impact of (1.5+1.5) kg (MEK + toluene) is about Rs.350/- whereas financial impact of 6.6 kg of hydrogen is about Rs.1200/-. Again, on the other side, additional financial impact against additional power consumption of 110 units in case of Table 5.5 over Table 5.6 is Rs.880/-. Hence, additional cost towards hydrogen in CIDW is offset by savings in solvent cost plus electricity cost. But in overall, there would be net energy cost savings towards less consumption of LP steam, circulating water and MP steam in CIDW operation than solvent dewaxing as evident from Table 5.5 and 5.6.

6 Tribology

In the preliminary section, Chapter 1, fundamentals along with chemistry on lubricants have been discussed, but it would remain incomplete without discussing lubrication mechanism. Lubrication mechanism is governed by the principles of rheology and tribology. Rheology covers the study on flow properties of the fluids, viz. fluid dynamics, particularly on the viscosity property of the fluid while in motion whereas tribology covers the friction between two rotating parts or between one rotating and one stationery parts of the machine with the presence of lubricant between these two parts and resulting effect on the machine parts with respect to wear.

Before discussing further on rheology or tribology, the lubrication mechanism as mentioned above should be explained. Lubrication can happen in three ways as below:

- Boundary lubrication.
- Mixed lubrication.
- Hydrodynamic lubrication.

Boundary lubrication is minimum lubrication, i.e. incomplete lubrication where two parts of the machine can come to direct contact to each other at one or two locations without lubrication film in between, mixed lubrication is the situation where there is inadequate lubrication film between two parts in one or two locations, and hydrodynamic lubrication is the scenario where there is adequate lubricant film all along the sliding parts of the machine. These three lubrication scenarios are best explained pictorially in Figure 6.1A.

The coefficient of friction is highest in boundary lubrication which decreases in mixed lubrication on an average throughout the film whereas it is moderate and low throughout the film in hydrodynamic lubrication as shown graphically by Stribeck curve as in Figure 6.1B.

There is another category of lubrication regime, i.e. elasto-hydrodynamic lubrication (EHL) regime. EHL is a typical lubrication regime for friction between parts (in sliding motion) having elastic contact under very high pressure, for example, ball bearings and gears. EHL is a scenario of hydrodynamic lubrication which takes into account elastic deflection of solid contacting surfaces.

DOI: 10.1201/9781003334422-6

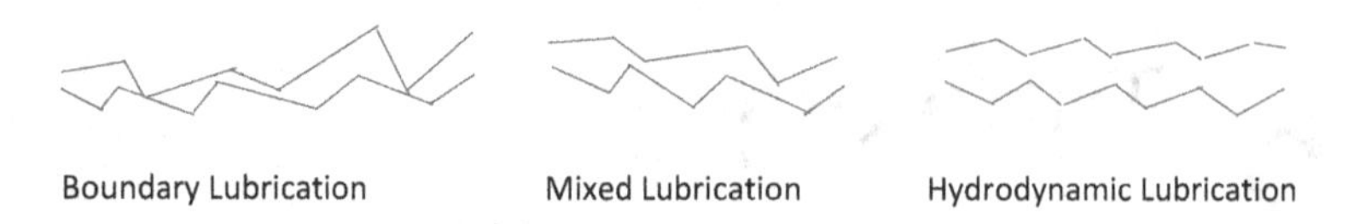

FIGURE 6.1A Different regime of lubrication.

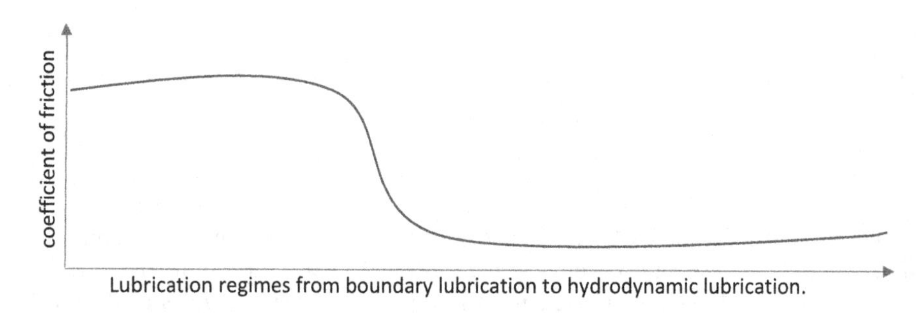

FIGURE 6.1B Coefficient of friction vs. lubrication regime.

In lubrication film between the two sliding surfaces, the film avoids the direct contact of the sliding parts thereby avoiding catastrophic friction leading severe degree of wear and damage, but the liquid film comes under a shear stress leading to shearing of the liquid film with the sliding / moving part with respect to its stationery part of the machine. In this process, the liquid viscosity should be adequate enough to reduce the shearing rate. The shearing stress and shear rate is governed by the following equation.

$$\text{Shear stress} = \text{Shear rate} \times \text{viscosity} \tag{6.1}$$

Note: when shearing of the liquid film is considered to happen, the liquid is called a non-Newtonian fluid, and hence, this viscosity values would follow the function of a non-Newtonian fluid as covered in detail on the subject, 'Rheology' in the literature. Hence, it is found that for fixed shear stress, if viscosity is high, then shearing of film would be low/nil which is good. In Chapter 2 the importance of viscosity and viscosity-temperature function like VI (viscosity index) has been discussed where other physico-chemical properties of lube base oils are also discussed as all these properties have direct or indirect effect on viscosity at the instant or its sustenance in the long term.

The objective should be to reduce the shear rate to reduce wear of the machine parts in relative sliding / movement. The study of this friction and wear is called tribology and without understanding lubrication mechanism as mentioned above it can't be evaluated. The engineering on this friction and wear covers all the following to inter-relate them in tribology:

- Friction
- Wear
- Lubrication mechanism

- Tribo chemistry
- Fluid dynamics
- Material science
- Life cycle assessment
- Machine design.

Tribo chemistry is the subject which covers development of various additive compounds which after being added to the lubricant improve the shear stability in the lubrication film thereby reducing friction and wear; these compounds are also called tribo additives. The viscosity modifier compounds which improves the viscosity index of the lubricant also indirectly contribute to the friction and wear.

Life cycle assessment is a balance between carbon footprint and carbon handprint; carbon footprint we know causes adverse impact of using the lubricant to the environment while carbon handprint is the advancement and development in the lubrication engineering which provides better technology(s) to ensure efficient lubrication while simultaneously ensuring extended drain interval of the lubricant.

The importance of tribology additives has been reflected in lubricant specification, IS-13656-2002 in sub-Section 3.1.3 of Chapter 3 where Table 1A and 5 reflect the importance of physico-chemical properties and Table 3 reflects the tribo properties.

Molybdenum di-thio carbamate (MoDTC) and molybdenum di-thio phosphate (MoDTP) are excellent tribology additives developed through tribo chemistry studies, i.e. they are highly soluble in the lubricant and are excellent antifriction, antiwear, and extreme pressure additives and are applied in lube oil formulation as discussed in Chapter 3. By reducing friction, it contributes to fuel saving by 3–5%[17].

Regarding tribo chemistry, i.e. chemistry of function of these additives, the explanation is [Mo-S additives decompose on contact with the hot surfaces and afford a mixed layer of MoS_2/iron sulfide/iron oxide which prevent metallic contact and bring the friction co-efficient values to a low level and exhibit higher load carrying capacity][17] thereby extending the drain interval of the lubricant.

7 Environment Impact and Mitigation in the Use of Lubricants

While automotive, industrial and domestic fuels made up from fossil fuels are responsible for air pollutions, lubricants made up from fossil fuels cause pollution to the soil only though 2T lubricant mixed with the burning fuel in two-wheelers cause air pollution.

As mentioned above, the lubricants are causing soil pollution during their disposal to soil after completion of their life cycle in the engines. The spent lubricants made from fossil fuels are not biodegradable and thus cause harm to the agriculture and ground water, also directly to human beings in handling the spent lubricant.

Among various categories of lubricants, disposal of spent hydraulic fluids contributes a large share of soil pollution because 90% of hydraulic fluids used in fire hazards applications are water based and thus cause soil pollution like affecting ground water.

In view of the increased awareness about environmental protection throughout the world, the hydraulic fluids having water content in their formulations have been focused to find out alternatives as these oils can leak into the soil, ocean, rivers, lakes or marshes and therefore can upset the ecological balance in such environment. Similarly, other lubricants like crankcase oils, gear box oils, brake oils, etc. also cause soil pollution during their disposal after completion of life cycles as these are not biodegradable though in their formulation water may not be there; but due to absence of water in their formulation, their impact on soil pollution is less than hydraulic fluids.

In view of the above, need to find alternative lubricants has found the priority in the advanced researches, and in this regard, while many biodegradable synthetic lubricants have been developed, various other bio lubricants from natural resources like edible or non-edible vegetable oils also have been developed.

Whether it is air pollution or soil pollution, contribution of the lubricants can be expressed in terms of carbon foot print. In the ecology balance cycle, if the lubricant has a net positive emission of carbon-dioxide, it is called carbon positive with respect to its carbon footprint; similarly, if it has net negative emission of carbon-dioxide, it is called carbon negative; also, if it has a net zero carbon-dioxide emission, it is called carbon neutral. From this angle, bio-lubricant made out of animal fats can't be called carbon neutral unlike bio-lubricant made out of vegetable oils which can be

DOI: 10.1201/9781003334422-7

called carbon neutral as because vegetable oils absorbs carbon-dioxide from atmosphere and thus it negates the effect of releasing carbon-dioxide to the atmosphere during the disposal of spent bio-lubricant to atmosphere or soil whereas animals release carbon-dioxide to atmosphere in the ecological cycle and thus bio-lubricant made out of it can't be called carbon neutral and instead such bio-lubricants are carbon positive but one can at least reduce demand for fossil fuel by using such lubricants.

As the bio lubricants are not in a position to replace full demand in lubricant industry, the other alternative like recycling of spent lubricant is also followed through established processes of recovery and thereafter doing further value additions like adding some proportion of fresh lubricant and/or adding some additives need to be done as per desired formulation.

Another alternative to fossil fuel-based lubricant is Ionic Liquid. First ionic liquid as lubricant was reported in 1914; subsequently, lot of researches were carried out and successfully used commercially as neat lubricant or as an additive to mineral oil-based lubricant or bio-lubricants.

In this chapter, the development and application of bio-lubricants and recovery of re-usable lube oil from spent lube oils have been discussed in detail along with a glimpse on ionic liquid.

7.1 BIO LUBRICANT

As discussed above, the bio-lubricant can be defined as the lubricant produced from bio source(s), but in true sense, bio-lubricant should be welcome if the same is made out of vegetable oil source(s) only or even if it is produced from synthetic components the lubricant should have biodegradability; for example, if the synthetic components are made out of animal fats or similar things and such lubricant is not used as 2T engine oils rather used in four-stroke engines which calls for disposal of the spent lubricant to the soil only and ultimately if the disposed spent lubricant is found to be biodegradable in the soil then such lubricant can be called low carbon positive because in one part of the carbon cycle carbon is released to atmosphere and in other part of the carbon cycle it is absorbed by the soil.

7.1.1 PERSPECTIVE

Biodegradability is the ability of the lubricant to be decomposed by microorganisms present in the soil, water and environment. In addition, lubricant itself or its degradation products should be non-toxic. It should have minimum or no adverse biochemical or physiological effect in living organisms.

As understood now, the biodegradability is the most important criterion for adapting it in the lubricant industry.

The criteria for biodegradability are determined by several types of laboratory screening tests; most widely used test is: CEC-L-33-94 which are explained in the tabulated form as Table 7.1.

A value of biodegradability of more than 70% is considered as adequate; the typical values of biodegradability of some lubricants are shown in the Table 7.2.

TABLE 7.1
Biodegradability Factor

Methods	Time (days)	Factor measured	Criterion
Readily Biodegradable:			
Modified AFNOR OECD 301A	28	Loss of dissolved organic carbon	>70%
Modified Sturn OECD 301A	28	Production of CO2	>60%
Modified MITI OECD 301D	28	Oxygen demand	>60%
Modified bottle OECD 301D	28	Oxygen demand	>60%
Modified OECD screening test	28	Loss of dissolved organic carbon	>70%
Inherent Biodegradable:			
Modified semi continuous activated sludge OECD 302A	>28	Loss of dissolved organic carbon	>20%
Zahn-Wellens EPMA test OECD 302B	28	Loss of dissolved organic carbon	>20%
Relatively Biodegradable (Primary Biodegradability):			
CEC-L-33-A-94	21	Loss of hydrocarbon infra-red bands at 2930 cm^{-1}	>67% and 80% (Blue Angel)

TABLE 7.2
Biodegradability Values

Type of Fluid	% Biodegradability as per CEC-L-33-A-94
Mineral oils	20–40
Vegetable oils	90–98
Monoesters	90–100
Diesters	75–100
Polyols	70–100
Complex Polyols	70–100
Poly Oleates	80–100
Phthalates	45–90
Di-merates	20–80
Tri-malliates	0–70
Pyro malliates	0–40

Typical criteria for determining toxicity of bio-lubricant or its biodegraded products are determined as described below:

- Aquatic and dermal toxicity – LD: 50 or less than 28/LG.
- Eye irritation – corneal opacity 2.0 or more.
- Conjunctivital redness (924–72 hours) – 2.5 or more.
- Skin irritation index (redness) – 2.0 or more.

TABLE 7.3
Impact of Aquatic Toxicity Level

Sl. no.	Impact	EC50 in mg/lit.
1.	Relatively harmless	>1000
2.	Practically non toxic	>100 to 1000
3.	Slightly toxic	>10 to 100
4.	Moderately toxic	>1 to 10
5.	Highly toxic	<1

TABLE 7.4A
Toxicity Values of Lubricants

Lubricant	WEN	WGK number	Classification
Vegetable oil	0–1.9	0	Non-hazardous to water
Lubricant base oils and white oils without additives	2–3.9	1	Slightly hazardous to water
Additive treated lubricants-engine oils and Industrial oils	4–5.9	2	Moderately hazardous to water
Additive treated water miscible lube oils (hydraulic oils), water miscible coolants	>6	3	Highly hazardous to water.

Aquatic toxicity levels are based on EC 50 values as shown in Table 7.3.

Typical toxicity values of some lubricants and additives are shown in Table 7.4A and 7.4B respectively.

7.1.2 Classification / Composition / Properties and Application / Commercialization

Commercially important classes of biodegradable lubricants are given as below:

- Vegetable oils and their chemically modified derivatives.
- Synthetic esters.
- Poly alkene glycols and its conjugates.

Vegetable oils are termed as glycerides of fatty acid from composition point view. We know that organic acid and alcohol combine in chemical reaction to form what is called an ester in general, but specifically, when the alcohol is a primary alcohol, viz. ethanol, propanol, butanol, etc. the trans esterified product is called ester, and in vegetable oil, as the alcohol is a tertiary alcohol, viz. glycerol, the product is called glyceride of fatty acid only instead of calling it as ester. Vegetable oils are long chain fatty acid combined with glycerol as naturally found. In the vegetable oil from a

TABLE 7.4B
Toxicity Values of Additives to Lubricants

	Typical LD values		Typical irritancy classification	
Additives	**Oral ratio mg/kg**	**Dermal (Rabbit) Mg/Kg**	**Eye**	**Skin**
Zinc alkyl dithio phosphate	>2000	>2000	Irritant	Irritant
Zinc alkaryl dithio phosphate	>5000	>2000	Not irritant	Not irritant
Calcium alkaryl sulfonate (long chain)	>5000	>3000	Not irritant	Not irritant
Calcium alkyl phenate (long chain)	>10000	>2000	Not irritant	Not irritant
Calcium alkyl phenate sulfide (long chain)	>5000	>2000 (Rat)	Not irritant	Not irritant
Polyolefin amide alkene amine	>10000	>2000	Not irritant	Not irritant
Polyolefin amide alkene amine borate	>2000	>3000	Not irritant	Not irritant
Polyolefin/Alkyl ester copolymer	>10000	>3000	Not irritant	Not irritant
Poly Alkyl methacrylate	>15000	>3000	Not irritant	Not irritant
Polyolefin	>2000	>3000	Not irritant	Not irritant

single source there are lots of glycerides with different molecular weight, i.e. with fatty acids of different carbon number whereas on the other side synthetic esters mean bio-lubricant made out synthetically, i.e. by getting the desired unsaturated long chain acid, say, oleic acid, linolic acid, etc. from the vegetable oil through hydrolysis chemical reaction thereafter carrying out trans esterification of this acid(s) with a primary alcohol, say, ethanol, propanol, or butanol, etc. to get the final ester as lubricant.

Similarly, when the alcohol used is glycol and the polymeric raw material is polyethylene or polypropylene or polybutylene, etc. then the combined product is called poly alkene glycol; in actuality, it is not produced like esterification process as mentioned above rather an alkene monomer is converted to epoxide followed by polymerization to get the polymer of epoxide followed by hydrolysis reaction to form the corresponding glycol as final product. The most common product is polyethylene glycol (PEG).

The chemical structures of ethanol, ethylene glycol, glycerol, oleic acid, polyethylene, and PEG are shown in the Figure 7.1.

Lubricity is obtained with increase in molecular weight. Hence, in the glycol group with polymeric chain of alkene (unsaturated alkyl group, say, polymeric chain of ethylene, propylene, etc.), lubricant is formed which is called a poly alkene glycol.

In glycerides of fatty acid, there would also be some double bonds, i.e. some degree of unsaturation in the alkyl group of fatty acid, and this acid replacing one hydrogen from CH_2 of glycerol form the ester called glyceride.

Vegetable oils and their chemically modified derivatives:

Typical characteristics of vegetable oils and their chemically modified derivatives as lubricants are as below:

i. Adequate viscosity range at ambient operating temperatures.
ii. Very high viscosity index (VI) which is highly preferable as lubricant.

C_2H_5OH	CH_2OH CH_2OH	CH_2OH $CHOH$ CH_2OH
Ethanol	Ethylene Glycol	Glycerol
(Primary Alcohol)	(Secondary Alcohol)	(Tertiary Alcohol)

$CH_3(CH_2)_7CH{=}CH(CH_2)_7COOH$
Oleic Acid

$\{-CH_2-CH_2-\}_n$
Polyethylene (PE)

$H-(O-CH_2-CH_2)_n-OH$
PEG

FIGURE 7.1 Chemical structure of different class of alcohols, oleic acid and PE.

iii. Good low temperature flow properties, i.e. very low pour point for oils and esters having higher degree of unsaturation. But it is poor in case of saturated fatty acid esters and glycerides.
iv. Poor thermos-oxidative stability, very poor in case of oils with high concentration of poly unsaturated fatty acid esters and glycerides.
v. Rapidly biodegradable and non-toxic to microorganisms.
vi. Good lubricity characteristics, i.e. lower friction, lower wear and higher welding loads.

Poor thermo-oxidative stability is improved by:

a. Partial hydrogenation of polyunsaturated fatty acid derivatives to monounsaturated derivatives.
b. Transesterification with oxo-alcohols of C_8 to C_{14} chain length.
c. Epoxidation of unsaturated double bonds.
d. Hydrogenation followed by acetylation of oils containing hydroxy and unsaturated fatty acid glycerides.
e. Formation of adducts by Diels Alder reaction with other short chain carboxylic acids and their anhydrides followed by hydrogenation.

Vegetable oils and their derivatives are commercially used as:

1) Hydraulic Fluids.
2) Machine tool Lubricants.
3) Total loss Lubricants.

Lubricants made out of vegetable oils and their derivatives have replaced 25% market of hydraulic fluids made out of mineral oil sources in Europe and 75% market in USA and Japan. Also, these bio lubricants have replaced 60% market of machine tool lubricants in Europe, Japan, and USA. Theses lubricants also replaced about 8% market share of total loss lubricant globally. Examples of 'The Total Loss Lubricants'

TABLE 7.5
Approximate Physico-chemical Properties of Vegetable Oils

Characteristics	Castor oil	Linseed oil	Rapeseed oil	Rice-bran oil	Sunflower oil
KV, cst @40°C	250	25	33	24	28
@100°C	20	7	8	6	7.5
VI	80–90	250	210	220	202–250
Iodine value	80–90	150–155	95–115	100–105	119–144
Saponi. value, mg KOH/g	170–190	170–180	170–180	180–185	180–195
TAN, mgKOH/g	<4	3.5	3.0	8.5	–
Pour point, °C	(–)9–(–)18	< (–)15	(–)5–(–)40	(–)3	(–)12

Notes:

1) As seen in the table, the acid numbers (TAN) are positive and are on the higher side. Hence, neutralization additive with good TBN (total base number) value, say, ZDP, should be added to make the lubricants marketable.

2) Iodine value, i.e. degree of unsaturation of Neem oil (not shown in the above table) is very good, i.e. on the lower side like castor oil while its other properties are also comparable except its VI which is on the lower side.

are: chain saw bar lubricants, mold release agents, two-stroke engine oils, wire rope lubricants.

Food processing and textile industries are rapidly changing over to vegetable oil-based lubricants.

It is worth mentioning about the physico-chemical properties of vegetable oils due to which these can be used as lubricants as mentioned above. The properties are shown in Table 7.5.

Regarding increased iodine values due to higher degrees of unsaturation in the lubricant, it is found that it is not an undesired property up to a certain limit, and also, this value can be reduced by addition of anti-oxidant additive in the lubricant blend formulation. An approximate estimate of degree of unsaturation in these vegetables is expressed as ratio of unsaturation with respect to carbon number in these oils are shown in Table 7.6.

From the limitation (as seen in Table 7.6) in the value of unsaturation in vegetable oils to a high degree, the anti-oxidant dosing sometimes may not be economical as well as technically feasible from compatibility point of view. Hence, an experiment was carried out for a suitable blending of vegetable oils with lube base oils based on fossil fuel as discussed in Chapter 2.

A case study on blend property evaluation:

With the above information, i.e. castor oil (a non-edible and abundantly available in India) having better qualities in overall basis, sample of castor oil and 800N lube base oil sample from fossil fuel source (HN lube base oil obtained from Haldia Refinery, Indian Oil Corporation Ltd. in 2001) were taken, analyzed and a blend between these two were done to find optimum blending as shown in Table 7.7 which shows an encouraging result.

TABLE 7.6
Degree of Unsaturation in the Vegetable Oils

characteristics	Castor oil	Linseed oil	Rapeseed oil	Rice-bran oil
C18:1	8.2	21	19	45.5
C18:2	3.6	17.4	14	27.7
C18:3	–	50.6	8.0	–

TABLE 7.7
A Case Study – Comparison of Properties of Vegetable, Mineral Based Lube Base Oil and Their Blend

characteristics	800N Base Oil (M)	Castor Oil (C)	Blend (M:C=50:50)
Density	0.89	0.96	0.925
Viscosity, KV, cst @100°C	14.0	17.0	15.5
@40°C	180	220	200
Viscosity Index (VI)	90	78.5	85
Iodine Value	0	85	42.5
Pour Point, °C	(–)3	(–)15	(–)6
Flash Point, °C	270	280	275
Saponification Value	1	180	90
TAN, mgKOH/gm	1	3	2

From Table 7.7, it is seen that its iodine value has come down to make anti-oxidant dosing technically feasible and economic also. However, as VI of castor oil was found low at 78.5, the VI of blended oil still low at 85 as compared to requirement of 90 minimum in the lubricant formulation; however, this shortfall can be made up by dosing VI improver in the final lubricant blend or otherwise ratio of castor oil can be reduced compromising iodine value to some extent if the additives dosing of VI improver and anti-oxidant need to be changed to get the right formulation. All other properties of the blend are found normal.

Synthetic Esters

Most important class of esters are:

- Dibasic acid esters.
- Neo pentyl polyesters.
- Complex esters.

Typical properties of these esters have been found to be as follows:

1) Right viscosity at ambient operating temperature.
2) Very high viscosity index (VI).
3) Excellent low temperature properties, i.e. having very low pour point.

4) Excellent hydrocarbon compatibility and solubility.
5) Normal to excellent thermos-oxidative stability.
6) Low volatility and hydraulic stability.
7) Poor elastomer compatibility.
8) Very high cost.
9) Good biodegradability.
10) Low eco-toxicity.

These lubricants are used in:

- Aircraft engine oils.
- Compressor oils.
- Precision instruments including medicinal instrument lubricants.
- Refrigeration oils where high quality oils are required.
- Hydraulic Fluids.

Approximate physico-chemical properties of some esters of MOBIL brands are shown along with brand names in the Table 7.8.

Poly alkene glycols:

These lubricants are water based and hydrocarbon soluble lubricants. Water based lubricants are poly ethylene and poly propylene glycols containing 200 to 1000 units of ethylene oxide or propylene oxide condensation products.

Hydrocarbon soluble poly alkene glycols are generally poly propylene or poly butylene glycols containing 100 to 400 units of propylene oxide or butylene oxide. The polyethylene glycol (PEG) in addition to use as lubricant is also used as raw material to produce further complex type of ester where the hydroxy groups are either etherified or esterified with branched chain isomeric acids or ethylene hexyl carboxylic acid. The final product lubricant falls under synthetic ester category of lubricant and are produced to use as lubricant for specific uses.

These lubricants are used as:

1) Hydraulic oils in earth moving machineries and hydraulic presses.
2) Metal working lubricants.
3) Rolling and mold release oils.

TABLE 7.8
Physico-Chemical Properties of Esters of MOBIL Brands

Brand	KV, cst@ 100°C	KV, cst@40°C	Pour point, °C	Flash point, °F
Esterex™ A32	2.8	9.5	<(−)85	405
Esterex™ A34	3.2	12.0	(−)76	390
Esterex™ A41	3.6	14.0	(−)71	448
Esterex™ A51	5.4	27.0	(−)71	477
Esterex™ NP343	4.3	19.0	(−)54	495
Esterex™ NP451	5.0	25.0	(-)76	491

TABLE 7.9
Approximate Physico-chemical Properties of Poly Ethylene Glycols (PEG)

Properties	PEG 400	PEG 200
Density, g/cc	1.12	1.124
Viscosity, cst @25°C	90	–
Viscosity, cst @100°C	–	4.3
Pour point, °C	4	(–)65
Boiling point, °C	>200	250
Molecular weight	360–440	190–210
Flash point, open cup, °C	–	190
P_H	5–7	4.5–7.5

Note: As seen in Table 7.9, [Sebacic is formed by caustic oxidation of natural fats and oils; Azelaic acid is formed by extracting Oleic acid from natural fats and oils followed by ozonolysis reaction][15], whereas adipic acid is relatively low mol. wt. compound with formula, $(CH_2)_4(COOH)_2$, dicarboxylic acid of butylene.

These lubricants are characterized by the following properties:

1) High viscosity adjustable by water to desired level.
2) High specific heat and latent heat for efficient heat removal.
3) High viscosity index.
4) Low temperature flow properties.
5) Poor hydrocarbon compatibility.
6) Good to moderate biodegradability.
7) Moderate toxicity.
8) Higher bio-accumulation.
9) Hydrocarbon soluble and compatible derivatives have high cost.

Regarding physico-chemical properties, the same for some poly-ethylene glycols are shown in Table 7.9 and poly ethylene glycol conjugate, i.e. the complex ester in Table 7.10.

It is also seen that biodegradable hydraulic fluid can be produced from all the above three above sources, i.e. vegetable oil, synthetic ester or poly alkene glycol.

Vegetable oil based hydraulic fluids possess excellent lubrication characteristics. The lubricant made out of rapeseed oil offers good corrosion protection and doesn't attack sealing materials, varnish and paints. It has some temperature limitations depending on extreme operating or climatic condition.

Synthetic ester based hydraulic fluids possess excellent lubricity at low as well as at high temperature and have ageing stability. But they are more expensive as compared to mineral oil based hydraulic fluids.

Poly ethylene based hydraulic fluids have wide applications in food industries, farming and construction equipment. These have temperature stability in the range of

TABLE 7.10
Physico-chemical Properties of Esters Made out of PEG

Terminal mono-alcohol	Diacid	Glycol	KV, cst @210°F	KV, cst @100°F	KV, cst @ (-)40°F	Dean Davis VI	Pour Point, °F	Flash Point, °F
2-ethylhexanol	Sebacate	PEG200	10.1	52.4	23,500	151	(−)60	480
2-ethylhexanol	Azelate	PEG200	8.8	43.8	18,500	153	<(−)70	460
2-ethylhexanol	Adipate	PEG200	7.2	35.7	21,500	153	(−)65	450
C_8-oxoalcohol	Sebacate	PEG200	11.0	57.7	25,900	151	(−)50	520
C_8-oxoalcohol	Azelate	PEG200	10.1	51.2	23,500	152	(−)70	485
C_8-oxoalcohol	Adipate	PEG200	7.1	34.8	18,700	155	(−)50	460
C_9-oxoalcohol	Sebacate	PEG200	13.6	77.1	–	146	(−)60	470
C_9-oxoalcohol	Adipate	PEG200	9.5	50.1	23,000	–	(−)50	435

(−)45 to 250°C and have biodegradability up to 99%. The oils change intervals are similar to those of mineral oils being 2000 hours or once a year.

7.1.3 Processes for Production

Processes for production of bio-lubricants are discussed in brief for each category as below:

1) Vegetable oils and fats:
   Like mineral oil-based lube base oils, vegetable oils are used as base oil for the production of bio-lubricant through blending process as discussed in Chapter 3 where different additives are added to these base oils based on formulation study to use the final lubricant for specific machine purposes as desired by OEM (original equipment manufacturer) of the machines. The base oil is selected from the Table 7.5 or extending it by including other vegetable oils as per machine requirement. It is seen from the Table 7.5 that all vegetable oils have TAN values at positive and the values are very high as compared to mineral oil-based base oil where TAN values are very low at one, and even in that case neutralizer additive, viz. ZDP (zinc alkyl di thiophosphate) is added to the base oil to get a TBN (total base number) value of 5. Hence, in case of producing bio-lubricant from vegetable oil, the dosing amount of ZDP would increase in the blend. As far as degree of unsaturation is concerned, that is also high in case of vegetable oils as reflected from the iodine values in Table 7.5; this is taken care of higher dosing of anti-oxidant compound in the blend to prevent gum formation, wear and corrosion in the machine.
2) Synthetic ester:
   Raw material for production of synthetic ester is also vegetable oils and fats but through chemical modification of these vegetable oils and fats, i.e. through desired chemical reactions as discussed below.

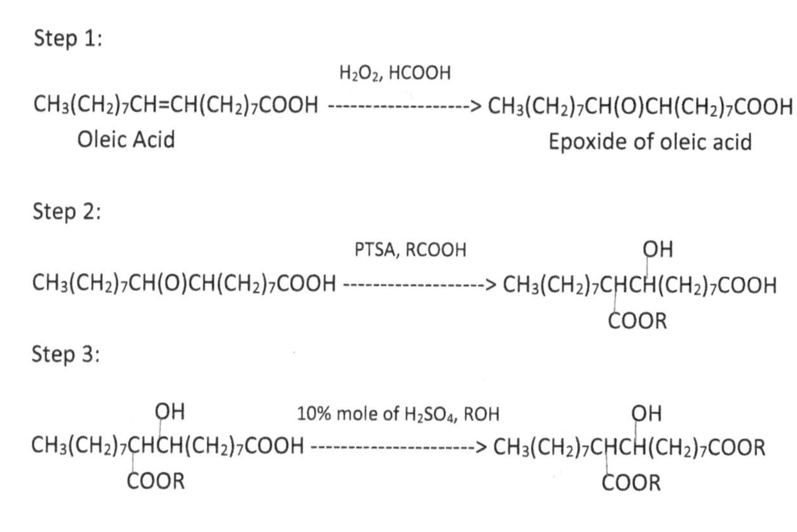

FIGURE 7.2 Reaction steps for production of synthetic ester.

First of all, the long chain fatty acid is produced through chemical reaction of decomposition/dehydration of vegetable oil/fat as per commercially established process. Through this reaction, various unsaturated fatty acids and glycerol are produced.

The fatty acid is then subsequently used as raw material for its chemical reactions in steps like epoxidation followed by epoxide ring opening reaction followed by esterification. An example with oleic acid as feed obtained from vegetable oil is discussed in 3 steps to get the ester as final lubricant as shown in Figure 7.2.[15]

3) Poly alkene glycol:
   To produce poly alkene glycol, alkene monomer, for example, ethylene, propylene, butylene, etc. are first converted to epoxide. Then the epoxide is polymerized where epoxide ring would remain but the alkene would be polymerized. Then the epoxide of polymer is hydrolyzed to get the corresponding poly alkene glycol. The reactions in 3 steps are shown in Figure 7.3.

7.2 DEVELOPMENT IN AUTOMOBILE INDUSTRY TO REDUCE LUBE OIL CONSUMPTION

Since the invention of the lube oil, there has been continuous development in the lube oil industry to extend the drain interval of lube oil from the machines thereby reducing lube oil consumption. This approach not only acts to control carbon footprint of the machines but also helps surviving in the competitive market to provide better quality lube oil to ensure enhanced machine life cycle as well as keeping sustainability in view.

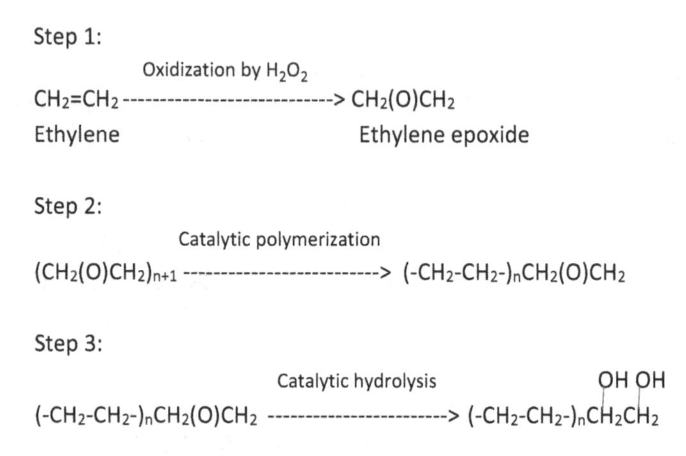

FIGURE 7.3 Reaction steps for production of poly-alkene glycol.

The chronological development in lube oil industry is summarized as below:

I. Synthesis of lube base oil from fossil fuel sources.
II. Development in the formulations of grease with base oils from fossil fuels.
III. Development of various additives and formulations with advancement in tribology researches to produce engine efficient lube oil through blending of these additives with the base oils and effective formulations.
IV. Single grade lube oil irrespective or summer or winter.
V. Multigrade lube oil adapting performance from summer to winter or vice versa.
VI. Development of bio-lubricant from vegetable oils and fats with objective to ensure biodegradability of the spent lube oil.
VII. Development of synthetic lube oil facilitating extended drain interval, ensuring better engine performances as well as ensuring eco-friendly disposal with respect to biodegradability of the spent lube oil.
VIII. Development of solid synthetic lubricant, viz. PTFE, graphite, nano-coating on the sliding surfaces of the machines thereby avoiding use of a liquid lubricant in high-speed machines, which are called self-lubricated machines.
IX. Conversion of two-wheeler engines from two-stroke to four-stroke preventing lube oil burning along with the fuel from the engines and instead facilitating lube oil disposal at a desired drain interval.
X. Development of synthetic ester-conjugates with vegetable oils and fats as raw material and synthesizing the substrate (intermediate) like oleic acid, linolic acid, etc. with minimum unsaturation (one for oleic acid and two for linolic acid) followed by producing next substrate as corresponding diacid and conjugating the same with a terminal mono-alcohol and poly ethylene glycol (PEG) to produce the final lubricant as complex ester. This type of esters is of premium qualities with improved thermal stability and other properties.

7.3 RECOVERY OF LUBE OIL FROM SPENT LUBE OIL

Used lube oil obtained during the disposal of the same from the machines in their drain intervals are having the lubrication properties intact except with the presence of contamination in it occurred from the wear of sliding parts of the machines and from surface of the chamber where it was kept. Hence, with the good waste management policies, these spent lube oils can be reclaimed and recycled after rectifying the same with respect to contamination dirt, water and other thermal / oxidization degradation products by processing the spent lube oil through standard / well established processes.

As per CONCAWE report, October, 2010, it is found that about 50% of these spent lube oils are collected in EU and about 36% is recycled after recovery of the quality lube oil from these spent lube oil. [As per US Environment protection agency (EPA) report also in 1988, it is found that about 59% of the spent lube oils enters the waste management system, 17% is dumped off by the industrial and non-industrial sources, 14% was disposed of by the individuals and 10% is disposed of by the generator][16].

Even now-a-days also, recycling industries have not come yet aggressively though the many reputed equipment manufacturers in the world accept the reclaimed oil after adding the necessary additives. As such, there are lots opportunities in this sector to increase reclaimed oil production to be incentivized by the respective government agencies to ensure a good waste management policies particularly for collection and marketing of the waste lube oil thereby discouraging the draining/burning of the spent lube oil by the stakeholders because in this process, the environment protection would be enhanced due to possible enhanced achievable reclaimed oils thereby reducing the fresh lube consumption and reducing the carbon footprint if the reclaim/recovery processes are well established.

This section discusses about various reclaiming processes as detailed below:

a) Conventional old method:
The spent lube oil is settled for hours and free water along with dirt is drained. Then the oil is heated above 100°C to evaporate out residual water and lighter hydrocarbon followed by cooling to ambient temperature. Then acid treatment is done to remove the asphaltene and residual dirt as sludge which is then filter pressed. The filtered oil is then bauxite clay treated, settled to drain out the clays to get the final product as lubricant with color improvement.

This method was being followed by small scale industries but now-a-days discouraged due to the environmental issues of disposal of acid sludge and spent clays in the process. The process is not economic also due to low scale of operation by the small-scale industries.

b) Viscolube process based on IFP technologies:
Viscolube, Italy together with IFP (France Institute of Petroleum) set up a reclaiming plant at Milan combining high vacuum high temperature distillation to recover lube distillates of different grades followed by propane de-asphalting of the distillation residue to get de-asphalted lube oil and finally

doing hydro finishing of all the lube oil grades obtained as above in IFP developed hydro finishing process to get the final marketable products as lube base oils.

In the Viscolube process, the spent lube oil first undergoes mild vacuum distillation to evaporate out the contaminated water in the spent lube oil. Then the oil is heated at a temperature of about 400°C at a high vacuum of about 700 mmHg to recover the lube distillates. The residue from the distillation is treated in liquid propane extraction unit to recover the raffinate, i.e. heavier lube fraction from this residue as de-asphalted lube oil where the extract obtained as by-product consists of the asphaltene including dirt, if any. The recovered lube distillates and de-asphalted oil are then treated in a IFP Hydro finishing unit in blocked out mode to get the different grades of lube base oils with improved color and oxidation stability.

c) The KTI process:
The Kinetic Technology International (KTI) developed this process based on thin film evaporation followed by hydrogenation. The steps in this process are as follows:
   - Removal of suspended solids, water and light ends through heating and settling.
   - Thin film vacuum distillation to produce lube distillate and residue containing metals, asphaltene and other polymerization products.
   - Hydrogenation to improve color and oxidation stability through reaction with nitrogen plus sulfur compounds and saturating the unsaturates with the hydrogen at high pressure and temperature to finally get the finished lube of improved color and oxidation stability respectively.
   - Vacuum distillation to get different fractions of lube base oil from the hydrogenated lube oil as mentioned above.

d) Mohawk process:
This process is same as KTI process except an additional chemical treatment is done after hydrogenation unit in the KTI process and operating the hydrogenation process at lower severity to ensure extended life of the catalyst in the hydrogenation process.

e) Phillips re-refined oil process (PROP):

This process consists of following steps:

- Mixing of the spent lube oil with di-ammonium phosphate and heating the mixture to separate out the metal contaminants. Then it is mixed with diatomaceous earth followed by filtration to remove metallic phosphate.
- Flash distillation to remove light hydrocarbons.
- Heating in a closed system in presence of hydrogen and also passing through a percolated bed of clay to ensure complete removal of color contributing compounds while passing it through the bed of clay if the same could not be hydrogenated completely beforehand. Texaco subsequently improvised this

TABLE 7.11
Tested Properties of Some Re-refined / Reclaimed Oils Samples

	A		B		C	
Re-refined oil	**Light**	**Heavy**	**Light**	**Heavy**	**Light**	**Heavy**
API gravity	**32**	**30**	**32**	**30.2**	**32.1**	**29.3**
Flash Pt., COC, °F	385	420	365	455	355	435
KV, cst @40°C	20.16	55.1	18.3	53.8	16.9	77.4
VI	85	102	86	85	92	102
Pour Pt., °F	10	15	10	10	20	5
ASTM color	<0.5	<1.0	<1.0	<1.5	<5.5	>8.0
Aromatics, %wt.	16.5	–	19.6	21.7	–	31.9

Source: [16].

process by developing nickel-molybdenum catalyst for efficient hydrotreatment thereby removing the subsequent clay treatment as mentioned above.

f) BETC process:
In this process, the spent lube oil is mild vacuum distilled to dehydrate the spent oil. Then the dehydrated oil undergoes liquid-liquid extraction with a mixed solvent of butyl alcohol, isopropyl alcohol and methyl ethyl ketone (MEK) to separate out the asphaltene and other polymerization products. The extracted oil is then treated through clay treatment or improved method like hydrogenation to get the improved color and oxidation stability in the final product of lube base oil.

g) UOP DCH process:
- Settling and filtration to remove dirt, water, and heavier asphaltene.
- Treating in hydrotreatment unit where in the catalytic reactor chemical reactions take place to get improved color, stability in the final product.
- Distillation to produce lube oils of different grades.

The typical physico-chemical properties of the reclaimed lube oils from different spent oils are shown in the Table 7.11.

All the reclaimed oils as shown in Table 7.11 fall into light category of lube oils as observed from the viscosity values.

Regarding economics of operation, both capital cost and operating cost of re-refining are higher than processes followed for fresh lube oil production; however, this information is based on the statistics collected from the operating reclaiming process units which generally operate at low scale as compared to lube units being operated at large scale by the corporate industries. But in any case, re-refining becomes economically attractive if the spent lube oils are available at cheaper prices in the market which can be ensured by improvement in the system of collection of the spent oils thus facilitating recovery plant operation at higher capacity.

7.4 IONIC LIQUID AS SUSTAINABLE LUBRICANT

Ionic liquid lubricant is a salt of complex organic compound with the complex compound is either cationic or anionic; the cations/anions are delocalized thus preventing them from condensing thereby facilitating the ionic liquids to remain in molten stage, i.e. to get very low pour point which is a desired property of a lubricant; not only this property, ionic liquids have all other properties like antifriction, anti-wear, low volatility, high flash point, higher viscosity and viscosity index, high anti-corrosion, high thermos-oxidative stability, and high biodegradability as well to use them as an ideal lubricant as an alternative to fossil fuel-based lubricants or bio-lubricants but ionic liquids are expensive at present, and hence, it is economic to use them as additive to other lubricants as mentioned above when it is observed that it is highly miscible in fossil fuel-based lubricants as well as in bio-lubricant. With further advancement in researches and technologies, in future it should be used as neat lubricant as it happens in every sector over a period of time as experienced by the mankind.

Present development of ionic liquid has also restricted to its use in low load machines which also should be overcome with further researches in this area.

8 World Lubricant Market and Major Operating Base Oil Refineries

World lubricant turnover is about 40 billion USD with CAGR of about 2 to 2.5%. We know while base oils are being produced mainly by the petroleum refineries, they sell it to lubricant marketing companies who use these base oils in their blending plants, apply their developed formulations and then produce the final finished lubricants to sell these to the bulk and retail customers for onward consumption. In some cases, these lubricant companies are integrated with the base oil refineries to have control in the upstream also.

Major lubricant marketing companies in the world are:

- Royal Dutch Shell
- BP
- Exxon-Mobil
- Chevron
- Petrobras SA
- Phillips 66
- Suncor Energy INC
- Valero Energy Corporation
- Ecopetrol SA
- Atlantis
- CNPC
- Sinopec
- Indian Oil Corporation Ltd.
- ONGC
- BPCL
- HPCL
- S Oil
- Eni SPA
- TOTAL
- Gulf Lubricant
- Castrol.

DOI: 10.1201/9781003334422-8

TABLE 8.1
List of Major Base Oil Refineries in the World

Refinery	Capacity, TMT/Yr. (approx.)	Thousand barrels/Day (approx.)
• North America:		
-Baytown Refinery, Texas	1650	32
-Baton Rouge, Lousiana	750	15
-Beaumont, Texas	570	11
-Houston, Texas	360	07
-Richmond, California	520	10
-Port Arthur, Texas	900	17
• Central and South America:		
-Rio De Janiero, Brasil	515	10
-Emmastadt, Netharland	465	9
• Europe:		
-CIS refineries in Europe (total)	8770	170
-Augusta, Italy	620	12
-Port Jerome, France	515	10
-Gonfreville, France	515	10
-Petit Couronne, France	470	9
-Leghorn, Italy	515	10
-All refineries, Germany	320	6
• Asia:		
-All refineries, China	2600	50
-All refineries, India	1600	30
• Middle East:		
-Duara, Iraq	200	4
-Teheran, Iran	200	4

Similarly, there are lots of petroleum refineries in the world producing base oils at total capacity of about one million barrels per day. The major base oil refineries are listed region wise in the world as shown in the Table 8.1.

References

1. *ASTM* D2270.
2. Exxon-Mobil Research Engineering (EMRE) comparison of LOBS (Lube Base Oil Stock) properties and their composition based on Middle East Crude sources.
3. Exxon-Mobil Research Engineering (EMRE) classification of LOBS (Lube Base Oil Stock) based on their compositions to match API based on Middle East Crude sources.
4. Licensed and commercialized by KBR, USA.
5. Avilino Sequeira, *Lubricant Base Oil and Wax Processing*, Aug. 9, 1994, CRC Press, pp. 6–7.
6. International Symposium on FUELS AND LUBRICANT, Dec. 8–10, 1997, New Delhi, Tata McGraw-Hill, pp. 173–174.
7. International Symposium on FUELS AND LUBRICANT, Dec. 8–10, 1997, New Delhi, Tata McGraw-Hill, p. 174.
8. International Symposium on FUELS AND LUBRICANT, Dec. 8–10, 1997, New Delhi, Tata McGraw-Hill, p. 30.
9. International Symposium on FUELS AND LUBRICANT, Dec. 8–10, 1997, New Delhi, Tata McGraw-Hill, p. 27.
10. International Symposium on FUELS AND LUBRICANT, Dec. 8–10, 1997, New Delhi, Tata McGraw-Hill, p. 54.
11. IS 13656: 2002-Internal Combustion Engine Crankcase oils (Diesel and Gasoline)-Specification (First Revision).
12. IS 14234: 1996-Internal Combustion Engine (Spark plug ignition with Gasoline as fuel)-Specification.
13. International Symposium on FUELS AND LUBRICANT, Dec. 8–10, 1997, New Delhi, Tata McGraw-Hill, p. 248.
14. Avilino Sequeira, *Lubricant Base Oil and Wax Processing*, p. 38.
15. Jumat Salimon, Nadia Salih, Emad Yousif, Raw materials, chemical modifications and environment benefits, *Eur. J. Lipid Sci. Technol.* 2010, 112, 519–530.
16. Avilino Sequeira, *Lubricant Base Oil and Wax Processing*, pp. 247–255.
17. International Symposium on FUELS AND LUBRICANT, Dec. 8–10, 1997, New Delhi, Tata McGraw-Hill, pp. 227–232.
18. B.K. Dutta, *Mass Transfer and Separation Process*, 5th edition, 2012; Membrane Separation Process, pp. 730–742.

SUGGESTED OTHER BOOKS TO STUDY

a. *Lubrication Fundamentals*, Revised and Expanded, Don M. Pirro, Webster Martin, and Daschner Ekkehard, 2016, Boca Raton, CRC Press.
b. Applications vary for dewaxing process over 10-year span, Smith F A, R W Bortz, *Oil and Gas Journal*, 1990, USA.
c. *Fundamentals of Petroleum and Petrochemical Engineering*, Uttam Ray Chaudhuri, Boca Raton, CRC Press, 2011.

Index

Note: Figures are indicated by *italics* and tables are in **bold type**. The acronyms "LOBS" and "NMPEU" are used for "lube oil base stock" and "n-methyl pyrrolidone extraction unit" respectively throughout the index.

A

ACEA (Association of European Engines Manufacturers) 106, 107, 112
additive classification: for automotive lubricants **81**; for industrial lubricants **81**
additive production 9–10
additive treated engine oils **169**
additives 89, 92, 93, 103, 104; anti-corrosion 95, 98, 100, 101, 102, 105; antifoam 6, 82, 95, 97, 98, 105; antifriction 165; anti-oxidant 84–85, 97, 105; anti-rust 95, 97, 98, 100, 101, 102, 105, 119; anti-wear 97, 98, 105, 119, 165; chemistry of 5–6; detergent 82; EP 98; extreme pressure 165; lubricant protective **81**; performance **81**; physico-chemical properties of 5–6; production technologies for 9–10; surface protective **81**; toxicity values of **170**; tribo 165; VII 83, 119
AEU (aromatic extraction of vacuum distillates) 29–40, *31*, *33*, *35*, **36**, **37**, **38**
agriculture spray oils 87, 104
aircraft engine oils 174
alcohols 6, 76, 77, 169–170, *171*, 171, **176**, 178, 181
aluminum rolling oils 86, 101–102
aniline point 115
anti-corrosion additives 95, 98, 100, 101, 102, 105
anti-corrosion properties 80, 96, 97, 109, 113, 118, 182
antifoam additives 6, 82, 95, 97, 98, 105
antifriction additives 165
anti-oxidant additives 84–85, 97, 105
anti-rust additives 95, 97, 98, 100, 101, 102, 105, 119
anti-wear additives 97, 98, 105, 119, 165
API (American Petroleum Institute) 18, **19**, 20, 73, 88, 106, 107, 112, 121
aquatic toxicity level, impact of **169**
aromatic extract 34, 64, *141*, 142
aromatic extracted distillate, solvent deoiling of 140–142, *141*, **141**
aromatic extraction 8, 10, 17–18, 23; solvent dewaxing of raffinate obtained from 40–55, *49*, **52**, **55**; of vacuum distillates (AEU) 29–40, *31*, *33*, *35*, **36**, **37**, **38**
aromatics 3–4, *4*, *7*, 8, 10, 15, *17*, 17, **19**, 20, **21**, 29, 32, 40, 64, 65, 66, 74, 75, 78, 85, 132, 142, 143, 149, **181**
ashless additive package vs. functions **93**
asphaltene 10, 14, 24, 28, 29, 124, 132, **139**, 140, 142, 144, 145, 148–149, 179, 180, 181
asphaltic compounds 12, 13
Association of European Engines Manufacturers (ACEA) 106, 107, 112
ASTM color **21**, **22**, 57, 80, **126**, **181**
auto transmission fluids 91–92, *91*
automobile industry, reducing lube oil consumption in 177–178, *178*
automobile lube oils 113, **114**
automobiles 15, 16, 78, 105, 106, 118
automotive greases 117, 118
automotive lubricants, additive classification for **81**
automotive lubricating oils 85, 87–92, **88**, **89**, *89*, *90*, *91*
axle oils 86, 99

B

base oil refineries 183, **184**
base oils 9, 17, 79, 80, 81, 85, 92, 94, 96, 97, 115, 178; chemistry of 2–4, *3*, *4*; refineries for **184**; see also lube base oils
bearing oils 86, 96–97, *96*, 99
benzene 41, 82
bio lubricants 166, 167–177, **168**, **169**, **170**, *171*, **172**, **173**, **174**, **175**, **176**, *177*, *178*
biodegradability factor **168**
biodegradability values **168**
blend property evaluation 172–173, **173**
blending 20, 108–112, **110**, *111*
boundary lubrication 82–83, 95–96, 163, *164*
brake oils 85, 91, 166
branding, of lube oils 112–113

C

candle industry 130
carbon Conradson residue (CCR) 14, **21**, **22**, 22–23, **26**, 28–29, **38**, 38–39, **52**, 53, 75, 80, 103, 107, 108, 113, 115
catalyst 56, 57, 58–59; and de-aromatization 68–69; and hydro finishing of deoiled wax 144; used in hydrotreating 66–67; used in

iso-dewaxing 67–68; used in reaction isotherm 67–68; system used for production of synthetic lubricant 76–77; and wax-hydro finishing unit (wax-HFU) 144, 145, 148–149
catalytic dewaxing unit 9, 18, 64, 65; utility consumption in **162**
catalytic iso-dewaxing: of vacuum distillate / raffinate 65–73, *71*, **71**, **72**; unit (CIDW) 9, 18–19, 25, 65, 69–73, *71*, **71**, **72**
catalytic processes 9, 151; for dewaxing 160–162, **161**, **162**
CCR *see* carbon Conradson residue (CCR)
cement plant greases 119
chain greases 119
chemistry 2–7, *3*, *4*, *6*, *7*; of additives 5–6; of base oils 2–4, *3*, *4*; tribo 165; of waxes 6–7, *6*, *7*
chilled solvent generation 47–48
CIDW (catalytic iso-dewaxing unit) 9, 18–19, 25, 65, 69–73, *71*, **71**, **72**
circulating hydraulic oil (anti-wear type 1) 94
circulating hydraulic oil (anti-wear type 2) 94
circulating hydraulic oil (R and O type) 94
classification: of additives for automotive lubricants **81**; of additives for industrial lubricants **81**; of greases 116, 117, **117**; of lube oil 85–87; of lube oil base stock (LOBS) 16–18, *17*, **19**, 20, 73; standards 106–107; of toxicity values **169**, **170**; of wax 123–124, 125
color: of LOBS 14, 16, 17, 18, 22, 23; of waxes 122, 124, **126**, **127**, **128**, **129**, 144
color stability, of LOBS 15, 16, 17, **21**, **22**, 22, 23
complex esters 173–174
compression ignition combustion engines 83, 85, 86, 87, **88**, 89, 92–93, 98, 105, 107
compressor oils 86, 97, 174; refrigeration 97
control valve hydraulic fluid 86, 94
conventional old method, of reclaiming lube oil from spent lube oil 179
corrosion inhibitor **81**, 82–83, 84, 91, 92, 100
cosmetics industry 125, 130
crankcase oils 85, 87, 89, *89*, *90*, 106, 166
crosshead and trunk piston type engine oils 87, 105
crude oil 2, 56, 76, 84, 121, 122; lube bearing 56, 74, 121; petroleum 4, 78, 10, 40; waxy 123, 142
crystallization 41, 42, 48, 53, 75, 132, 134–135, 136; filtration of wax-oil slurry after 42–44; temperature of 42

D

DAO (de-asphalted oil) 17–18, 22, 23–24, 25, **26**, 27, 28, 29, 30, **38**, 62, 140, 180
de-aromatization 65, 68–69
de-asphalted oil (DAO) 17–18, 22, 23–24, 25, **26**, 27, 28, 29, 30, **38**, 62, 140, 180
de-emulsifier 6, **81**, 84, **93**
denitrification 4, 56, 57, 60, 64, 66, 67, 145
deoiled wax *133*, 133, 136, 142, 143, 154, 157; hydro finishing of 6, *6*, 143–149, *147*, **147**, **148**; typical properties of 138, **139**, 140–141, *141*
deoiling *see* solvent deoiling
deoiling temperature 132–133, *133*, 157
deoiling unit, for paraffin wax production **153**, **154**, 154, 156, 157, 159, 160
desulfurization 56, 57, 60, 64–65, 67, 144, 145
detergent 6, 80, **81**, 83, 89, 91, 92, 93, 98, 113
detergent additives 82
dewaxed oil (DWO) 8, 23, 40, 75, 133, 140; desired properties of **52**, 52–54; in dewaxing and deoiling process 154, 157, 159, *159*, 160; hydro-finishing of 55–63, *61*, **62**, **63**
dewaxing: catalytic 9, 18, 20, **21**, 64, 65, 151, 160, **162**; solvent *see* solvent dewaxing
dibasic acid esters 173–174
diesel 9, 64–65, 70, *71*, 83
diesel engines 83, 85, 86, 87, **88**, 89, 92–93, 98, 105, 107
dispersant 6, **81**, 81–82, 92, **93**, 94
dispersion 80, 81–82, 89, 91, 92, 109, 113
drop melting point (DMP) 2, 7, 10, 115, 122, 124, **126–129**, 130, **139**, 143, 145
drop point 113, 115, 116, **117**, 118–119
DWO *see* dewaxed oil (DWO)
DWO yield 42, 43, 51, 53, 54, 142
DWO-solvent recovery circuit 45–46, 47, 48

E

elasto-hydrodynamic lubrication (EHL) 163
energy consumption 23, 58, 134, 151, 153, 154–155, 156, **156**, 160
engine oils: additive treated **169**; aircraft 174; crosshead and trunk piston type 87, 105; four-stroke 167; gas 85, 91; stationary 86, 98; two-stroke 85, 89–91, *89*, *90*, 172
engine test performance standards 106, 107
environment impact, of lubricants 166–182, **168**, **169**, **170**, *171*, **172**, **173**, **174**, **175**, **176**, *177*, *178*, **181**
EP (extreme pressure) additives 98, 165
esters, synthetic 20, 76, 169, 170, 173–174
explosive industry 131
extraction: aromatic *see* aromatic extraction; NMP (n-methyl pyrrolidone) 36, **37**
extraction temperature 23, 24, 32, *33*, 33
extreme pressure (EP) additives 98, 165

F

FDA (Food and Drug Administration) test 104, 123

feedstock 7–8, 9, 10, 14, 17, 18, 20, 21, 23, 24, 29, 30, 33, 34, 35, 42, 57, 62, 64, 65, 67, 68, 72–73, 74, 75, 96, 103, 121, 132, 133, 140, 143, 154
FEU (furfural extraction unit) 23, 29–35, *31*, *33*, *35*, **36**, 36, 40
filter circuit 50–52
filtration 45, 48, 50–52, 53, 75, 135–136, *135*, 142, 157, 180, 181; temperature 10, 132–133, 159; of wax-oil slurry after crystallization 42–44
fire resistant hydraulic fluids (water glycol type) 95
fire resistant hydraulic oils (dilute emulsion type) 95
flash point 15–16, 20, **21**, **22**, **26**, 182; of aluminum rolling oils 101; of engine oils (crosshead and trunk piston type) 105; of esters of MOBIL brands **174**; of finished wax 146; of four-stroke crankcase oil *89*; of gear oil *91*; of hydraulic oil *95*; of industrial gear oil *96*; of LOBS 60; of methyl ethyl ketone (MEK) **134**; of metal working oils 100; of methyl iso-butyl ketone (MIBK) **134**; of neat cutting oils 101; of neat honing oils 101; of paraffin wax **127–129**; of poly ethylene glycols (PEG) **175**, **176**; of quenching oils 102; of refrigeration compressor oil *98*; of soluble cutting oils 100, **100**; of spark erosion fluids 101; of toluene **134**; of turbine oil *94*; of two-stroke crankcase oil *90*
fluid dynamics 23, 44, 163, 165
Food and Drug Administration (FDA) test 104, 123
foots oil 121, *133*, 133, *137*, 138, **139**, 140–141, *141*, 142, 154, 159, 160
four-stroke crankcase oils 89, *89*
four-stroke engine oils 167
four-stroke engines 87, 167
friction improver 80, 91
friction modifiers 5, **81**, 88
furfural 8, 23; degradability of 34
furfural extraction unit (FEU) 23, 29–35, *31*, *33*, *35*, **36**, 36, 40

G

gas engine oils 85, 91
gasoline engines *see* four-stroke engines; two-stroke engines
gear oils 85, 86, 87, 90, 91, *91*, 96, 105
general equipment and machinery greases 118–119
graphited greases 119
grease 113–119, **114**, **117**
Group-I LOBS 14, 17, 21–29, **22**, *26*, **26**, **27**
Group-II LOBS 14, 64, 65, 73
Group-III LOBS 17, 18–19, 63–73, *71*, **71**, **72**
Group-IV base oil *see* lube oil
Group-IV LOBS 19, 20, 75, 78
Group-V LOBS 18, 19, 20

H

heat exchangers 23, 24–25, 30, 31, 36, 37, 44, 45, 46, 47, 48, *49*, 59–60, 63, 69, 70, 72, 145–146, 148, 151, 152, *153*, 160
heat integration 151, 152–153, *153*, **153**, **154**
heat transfer oils 103
hydraulic fluids 85–86, 94–95, 166, 171–172, 174, 175–176
hydraulic oils 86, 87, 94–95, *95*, 105, **169**, 174–175
hydro finishing, of deoiled wax 143–149, *147*, **147**, **148**
hydrocarbons 2–4, *3*, 9, 10, 12, 13, 17, 57, 68, 180
hydrocracker bottom 19; solvent deoiling of 143; solvent dewaxing of 75
hydrocracking 65–66, 68, 143; and solvent dewaxing 73–75
hydrodynamic lubrication 163, *164*
hydro-finishing of dewaxed oil (DWO) 55–63, *61*, **62**, **63**
hydro-finishing unit (HFU) 9, 18, 23, 25, 37, **38**, 56–63, *61*, **62**, **63**, 124, 138, 140, 143–144, 144–145, 146
hydrogen purity 57–58, 66
hydrotreating 65–66, 66–67, 68, 69, 73, 74–75
hydrotreating unit 57, 63, 148

I

IC (internal combustion) engines *see* four-stroke engines; two-stroke engines
ILSAC (International Lubricant Standards and Approval Committee) 106, 107, 112
industrial gear oil 94, 96, *96*
industrial greases 117, 118–119
industrial lubricants **81**, 93, 99, 113
industrial lubricating oils 85–86, 93
industrial machines 16, 80, 95, 106, 112, 118
industrial specialty oils 87, 102
industrial white oils 87, 104
internal combustion (IC) engines *see* four-stroke engines; two-stroke engines
International Lubricant Standards and Approval Committee (ILSAC) 106, 107, 112
interphase in extractor and controlling extraction, principles of 34–35, *35*
investment cost: of CIDW unit 72–73; of furfural extraction unit (FEU) 32–34, *33*, 37–40, **38**; of hydro-finishing unit (HFU) 63; of NMPEU 37–40, **38**; of propane de-asphalting unit

(PDAU) 27–29; of solvent dewaxing unit (SDU) 55; of solvent deoiling unit (SDU-2) 139–140; of wax-hydro finishing unit (wax-HFU) 148–149
ionic liquid 167; as sustainable lubricant 182
iso-dewaxing (ISDW) 67–68, 69, **72**; *see also* catalytic iso-dewaxing
iso-paraffin *3*, 3, 4, 6, *6*, *7*, 12, 13, 14, *17*, 64, 67, 68, 74, 143

J

JASO (Japan Automobiles Standards Organization) 106, 107, 112

K

Kinetic Technology International (KTI) process 180

L

life cycle assessment 165
lube base oil production 8–9, 121, 154
lube base oils 8, 9, 84, 122, 164, 172, 176, 180
lube bearing crude oil 56, 74, 121
lube oil 79; applications of 87–106, **88**, **89**, *89*, *90*, *91*, **93**, *94*, *95*, *96*, *98*, **100**; branding of 112–113; classifications of 85–87; compositions of 87–106, **88**, **89**, *89*, *90*, *91*, **93**, *94*, *95*, *96*, *98*, **100**; marketing of 106–107, 112–113; production processes for 108–112, **110**, *111*; properties of **21**, **22**, 79–85, **81**; recovery from spent lube oil 179; significance of 79–85, **81**
lube oil base stock (LOBS) 11, 16; API classifications of **19**; API groups of **19**; application of 77–78; carbon Conradson residue (CCR) 14; chemical composition of 16–20, *17*, 18–19, **19**; classifications of 16–20, *17*, **19**; color 14; color stability 15; flash point 15–16; grades of 11, 16, 17–18, **21**, **22**, **71**, 77; Group-I 14, 17, 21–63, **22**, *26*, **26**, **27**, *31*, *33*, *35*, **36**, **37**, **38**, *49*, **52**, **55**, *61*, **62**, **63**; Group-II 63–77, *71*, **71**, **72**; Group-III 63–77, *71*, **71**, **72**; Group-IV 19, 20, 75, 78; Group-V 76; manufacturing technologies for 21–77, **22**, *26*, **26**, **27**, *31*, *33*, *35*, **36**, **37**, **38**, *49*, **52**, **55**, *61*, **62**, **63**, *71*, **71**, **72**; market specifications of 20–21, **21**; neutral (N) 17, 18, 20, **21**; oxidation stability 15; physico-chemical properties of 11–16, **12**, 18; pour point 13–14; solvent neutral (SN) 17, 18, 20, 21, **36**, **37**, **52**, **62**, 78; standards of 20–21, **21**; viscosity 11–12, **12**; viscosity index 12–13
lube oil marketing specifications 106–107
lube oil production / blending, processes followed in 108–112, **110**, *111*
lubrex cup greases 119
lubricant marketing companies 100, 113, 183
lubricant protective additives **81**
lubricants: additives to *see* additives; automotive **81**; bio *see* bio lubricants; definition of 1; environment impact of 166–182, **168**, **169**, **170**, *171*, **172**, **173**, **174**, **175**, **176**, *177*, *178*, **181**; machine tool 171–172; metal working 86, 99–100, 174; mitigation in use of 166–182, **168**, **169**, **170**, *171*, **172**, **173**, **174**, **175**, **176**, *177*, *178*, **181**; sustainable 182; synthetic 2, 18, 19, 75, 76–77, 166, 178; total loss 171–172; toxicity values of **169**
lubrication: boundary 82–83, 95–96, 163, *164*; elasto-hydrodynamic (EHL) 163; mixed 163, *164*; hydrodynamic 163, *164*; Webster's model of 80

M

machine tool lubricants 171–172
machine tool way oils 86, 98
machinery oils 86, 95
manufacturing technologies: for lube oil base stock (LOBS) 21–77, **22**, *26*, **26**, **27**, *31*, *33*, *35*, **36**, **37**, **38**, *49*, **52**, **55**, *61*, **62**, **63**, *71*, **71**, **72**; for wax 131–149, *133*, **134**, *135*, *137*, **138**, **139**, *141*, **141**, *147*, **147**, **148**
marine lubricating oils 87, 105
marine steam cylinder oils 87, 105–106
marine turbine oils 87, 105
marketing specifications: of lube oil 106–107; of wax 124–125
mar-quenching oils 102
MCW (micro-crystalline wax) 6, *7*, 7, 51, 123–124, 125, **129**, **130**, 130, 131–132, 137, **138**, **139**, 139, 140, **141**, 142, 143, 145, **147**, 148, 149, 154, 155, 156, **156**
MEK (methyl ethyl ketone) 10, 41, 46, 54, **55**, 133–134, **134**, 141, **141**, **156**, 157, 159, **161**, 162, 181
membrane separation 151, 152, 156–160, **156**, *158*, *159*
membrane separation unit 159, *159*
metal working lubricants 86, 99–100, 174
metal working oils 86, 99–100
metallurgy: of catalytic iso-dewaxing unit (CIDW) 72–73; of furfural extraction unit (FEU) 32–34, *33*, 37–40, **38**; of hydro-finishing unit (HFU) 63; of NMPEU 37–40, **38**; of propane de-asphalting unit (PDAU) 27–29; of solvent dewaxing unit (SDU) 55; of solvent deoiling unit (SDU-2) 139–140; of wax-hydro finishing unit (wax-HFU) 148–149
method of reaction 77

methyl ethyl ketone *see* MEK (methyl ethyl ketone)
MIBK (methyl iso-butyl ketone) 10, 133–134, **134**, 138, **139**, 141, **141**, **153**, **154**, **156**, 157, 159
micro-crystalline wax *see* MCW (micro-crystalline wax)
mineral oils 86, **100**, 105–106, **168**, 176
miscibility temperature 32, *33*, 44, 133
mitigation in use, of lubricants 166–182, **168**, **169**, **170**, *171*, **172**, **173**, **174**, **175**, **176**, *177*, *178*, **181**
mixed lubrication 163, *164*
Mohawk process, for reclaiming lube oil from spent lube oil 180
molding oils 104
Morgan bearing oils 86, 96–97

N

N (neutral) LOBS 17, 18, 20, **21**
naphthenes *3*, 3, 4, **19**, 23–24, 64, 65, 74
National Lubricating Grease Institute (NLGI) grades of grease 116, **117**
natural waxes 121–122, 125, 130, 149
neat cutting oils 86, 101
neat honing oils 86, 101
neo pentyl polyesters 173–174
neutral (N) LOBS 17, 18, 20, **21**
Newtonian fluids 115–116
NLGI (National Lubricating Grease Institute) grades of grease 116, **117**
NMP (n-methyl pyrrolidone) 8, 23, 36, 37, 40
NMP extraction unit (NMPEU) 23, 25, 36–40, **37**, **38**, 55, 56, 65–70, *71*, **71**, **72**
non-Newtonian fluid 115–116, 164
n-paraffin 2, *3*, 3, 4, 6, *6*, *7*, 7, 9, *17*, 67–68, 143, 149

O

oil refineries 183, **184**
oily aromatics *see* aromatic extract
oily wax *see* slack wax
oily wax from tank bottom, solvent deoiling of 142–143
olefin saturation 8–9, 57, 66
oleic acid 170, *171*, *177*, 177, 178
open gear compounds 99
oxidation stability: of base oils 2, 3, 4; of LOBS 15

P

packaging industry 125
paints and polish industry 131
PAO (poly alpha olefin) *see* lube oil
paraffin wax *6*, 7, 123–124, 125, **126–128**, 130–131, 132, **147**; production of **138**, **139**, 140, 141, **141**, **153**, **154**, 156, **156**
PDAU (propane de-asphalting unit) 18, 22, 23–29, *26*, **26**, **27**
PE wax (polyethylene wax) 122, 149, *171*
PEG (poly ethylene glycols) 170, *171*, 174, **175**, 175, **176**, 178
penetration number 115, 116
performance additives **81**
permeation through membrane, theoretical principles for 158–160, *159*
petroleum crude oil 4, 78, 10, 40
petroleum jelly industry 131
pharmaceutical industry 125
Phillips re-refined oil process (PROP) 180–181, **181**
physico-chemical properties 2; of additives 5–6; of esters **174**, **176**; of four-stroke crankcase oils 89, *89*; of gear oils 91, *91*; of hydraulic oils 95, *95*; of industrial gear oil 96, *96*; of LOBS 11–16, **12**, 18; of lube oil **21**, **22**, 80; of micro-crystalline wax (MCW) 7; of paraffin wax 7; of poly ethylene glycols (PEG) **175**, 175; of reclaimed lube oils **181**, 181; of refrigeration compressor oil 97, *98*; standards 106, 107; of turbine oils *94*, 94; of two-stroke engine oils 90, *90*, 91; of vegetable oils **172**, 172
pinch technology 151, 152–153, *153*, **153**, **154**
pneumatic tool oils 86, 98
polish and paints industry 131
poly alkene glycols and its conjugates 169, 174
poly alpha olefin (PAO) *see* lube oil
poly ethylene glycols (PEG) 170, *171*, 174, **175**, 175, **176**, 178
polyethylene wax (PE wax) 122, 149, *171*
pour point improver (PPI) 80–81, 84
pressure characterization 80, 85
PROP (Phillips re-refined oil process) 180–181, **181**
propane de-asphalting unit (PDAU) 18, 22, 23–29, *26*, **26**, **27**
PVC pipes and fittings industry 130

Q

quenching oils 87, 102

R

radiator coolant 85, 92
raffinate 140–142, *141*, **141**
rail road oils 85, 92–93
reaction constraints: in de-aromatization reactor 69; hydrotreating 67; in iso-dewaxing 68; in wax isomerization 68

reaction isotherm, catalyst used in 67–68
reaction temperature 57, 58, 67, 68, 69, 70, 72–73
reclaimed lube oils **181**, 181
reclaiming processes 179–181, 181
refractive index (RI) 35, **38**, 38, 39–40
refrigeration 41, 50, 55, 134–135, 136, 140, 157
refrigeration compressor oils 86, 93, 97, *98*
residence time 9–10, 42, 57, 58, 74, 108, 111
rheology 5, 125, 131, 163
RI (refractive index) 35, **38**, 38, 39–40
rotary vacuum filtration *135*
rubber components and rubber products industry 131
rubber process oils 103
rust inhibitor 6, 84
rust preventive oils 103

S

SDU *see* solvent dewaxing unit (SDU)
selectivity 32–33, *33*, 36, 75, 160
shear stability index (SSI) 84
shock absorber oils 85, 92
simultaneous dewaxing and deoiling 151, 154–156, **155**
slack wax **52**, 52, 54, 64, 74, 121, 132, 140, *141*, 142–143, 149, 152, 154–155, **155**, *158*
SN (solvent neutral) LOBS 17, 18, 20, 21, **36**, **37**, **52**, **62**, 78
solubilizing property 115
soluble cutting oils 86, 100–101
solvent and product recovery systems 136–137, *137*, **138**
solvent deoiling: of aromatic extracted distillate 140–142, *141*, **141**; of hydrocracker bottom 143; of oily wax from tank bottom 142–143; of vacuum distillate *133*
solvent deoiling unit 121, 132, 133, 137, *137*, **138**, 138–139, 140, 152
solvent dewaxing 6, 10, 36, 49; and hydrocracking 73–75; of hydrotreater / hydrocracker bottom 75; of raffinate obtained from aromatic extraction 40–44; utility consumption in 160, **161**
solvent dewaxing unit (SDU) 8, 17–18, 19, 23, 25, 40, 41, 42, 75, 121, 132, 133, 139, 140, 152, 153, 157, **162**; process description of 44–55, *49*, **52**, **55**
solvent dewaxing with membranes *158*
solvent dilution 41, 42, 43, 133, 134
solvent drying column 30–32, 48, 135, 136
solvent neutral (SN) LOBS 17, 18, 20, 21, **36**, **37**, **52**, **62**, 78
solvent power 32, *33*, 33, 36, 45, 104, 133
solvent ratio 24, 28, 30, 32, *33*, 33, **36**, **37**, 39, 42, **52**, **138**, **155**, 156, 157
solvents 41, 48, 54, 157; *see also* benzene; toluene
space velocity 57, 58, **62**, 62, 67, 68, **147**
spark erosion fluids 86, 101
spark ignition combustion engines *see* four-stroke engines; two-stroke engines
special purpose hydraulic oils (anti- wear type) 86, 94–95
spent lube oil, recovery of lube oil from 179
spindle oils 86, 95
SSI (shear stability index) 84
stationary diesel engines 98
stationary engine oils 86, 98
steam cylinder oils 86, 98; marine 87, 105–106
steel rolling oils 86, 102
stern tube oils 87, 106
strip grinding oils 86, 101
surface protective additives **81**
sustainable lubricant, ionic liquid as 182
synthetic esters 20, 76, 169, 170, 173–174
synthetic lubes *see* Group-IV LOBS; Group-V LOBS
synthetic lubricant 2, 18, 19, 75, 76–77, 166, 178
synthetic soluble cutting oils 86, 101
synthetic wax 121, 122, 131, 149

T

tarpaulin industry 130
temperature gradient in the extractor 35
testing standards 88, **89**
textile oils (biodegradable) 86, 96
textile oils (scourable type) 86, 95–96
thermal stability 1, 13, 80, 85, 89, 91, 95, 96, 97, 98, 99, 102, 105, 113, 118, 160, 178
thermo-oxidative stability 171
toluene 10, 41, 46, 54, **55**, 76, 133, **134**, 134, 141, **141**, **156**, 157, 159, **161**, 162, *177*
total loss lubricants 171–172
toxicity 49, 168, **169**, 169, **170**, 174, 175
tribo additives 165
tribo chemistry 165
tribology 5, 79, 82, 83, 84, 163–165, *164*
tribology additives 165
turbine oils 85, 93, *94*, 94, 97; marine 87, 105
two-stroke crankcase oil 90, *90*
two-stroke engine oils 85, 89–91, *89*, *90*, 172
two-stroke engines 87, 89, 107
tyre industry 131

U

unsaturates 17, 32, 56, 180
utility consumption: in catalytic dewaxing 160–162, **161**, **162**; in catalytic iso-dewaxing unit (CIDW) 70, **72**; through conventional processes **161**; of deoiling unit for paraffin wax

production with application of pinch technology **153**, **154**; in furfural extraction unit (FEU) 36, **37**; in hydro-finishing unit (HFU) **63**, 63; in NMPEU 36–37, **37**; in propane de-asphalting unit (PDAU) **27**, 27; in solvent dewaxing unit (SDU) 54–55, **55**; in solvent deoiling unit (SDU-2) 138–139, **139**; in solvent dewaxing 160, **161**; in wax-hydro finishing unit (wax-HFU) **148**, 148

V

vacuum distillates 18, 19, 21, 23; aromatic extraction of (AEU) 29–40, *31*, *33*, *35*, **36**, **37**, **38**
vacuum generation circuit, in filtration section 50
vacuum pump oils 86, 98
vegetable fats 176, 178
vegetable oils 9, 20, 76, 166–167, **168**, 173, 176, 178; and their chemically modified derivatives 169, 170–172, *171*, **172**
VGC (viscosity-gravity constant) 115
VI (viscosity index) 1, 8, 9, 12–13, 80, 92, 164, 165, 170, **173**, 173, 175, 182
VII (viscosity index improver) 80, **81**, 81, 83–84, 88, 91, 178
VII additives 83, 119
viscolube process, for reclaiming lube oil from spent lube oil 179–180
viscosity: of lube oil base stock (LOBS) 11–12, **12**; of waxes 123
viscosity index (VI) 1, 8, 9, 12–13, 80, 92, 164, 165, 170, **173**, 173, 175, 182
viscosity index improver (VII) 80, **81**, 81, 83–84, 88, 91, 119, 178
viscosity modifying polymer 5
viscosity vs. viscosity function **110**
viscosity-gravity constant (VGC) 115

W

wax: application of 125–131, **126–130**; chemistry of 6–7, *6*, *7*; classification of 123–124, 125; color of 122, 124, **126**, **127**, **128**, **129**, 144; derivatives of 149–150; drop melting point of 122; Food and Drug Administration (FDA) test for 123; isomerization of 68; manufacturing technologies for 131–149, *133*, **134**, *135*, *137*, **138**, **139**, *141*, **141**, *147*, **147**, **148**; marketing specifications of 124–125; natural 121–122, 125, 130, 149; oil content of 122; production of *see* wax production; properties of 122–123; significance of 122–123; synthetic 121, 122, 131, 149; viscosity of 123
wax match industry 131
wax-HFU (wax-hydro finishing unit) 143, 144–149, *147*, **147**, **148**
wax-oil slurry, filtration after crystallization of 42–44
wax-solvent recovery circuit 47, 48, *158*
waxy crude oil 123, 142
Webster's lubrication and cooling model 80
world lubricant market 183, **184**

Printed in the United States
by Baker & Taylor Publisher Services